Shapes
of
Time

KENNETH J. McNAMARA

Shapes
of
Time

The Evolution of Growth
and Development

The Johns Hopkins University Press   Baltimore and London

Printed in the United States of America on acid-free recycled paper
06 05 04 03 02 01 00 99 98 97   5 4 3 2

The Johns Hopkins University Press
2715 North Charles Street
Baltimore, Maryland 21218-4319

Library of Congress Cataloging-in-Publication Data
will be found at the end of this book.

A catalog record for this book is available from the British Library.

ISBN 0-8018-5571-3

For Susie, Jamie, Katie, and Tim

An empty book is like an Infant's

Soul, in which anything may be

written, it is capable of all things

but containeth nothing. I have

a mind to fill this with profitable

wonders . . .

Things strange yet common,

most high, yet plain: infinitely

profitable, but not esteemed;

Truths you love, but know not.

Thomas Traherne
Adapted by Gerald Finzi
from *Centuries of Meditation* I 1.2.3

Contents

Acknowledgments

MY AIM WITH THIS BOOK IS TO TELL THE STORY OF HOW CHANGES IN THE rates and timing of development provide the all-important link between, at one end, genetics and, at the other, natural selection. It is about how shapes evolve; how sizes evolve; how life history strategies, such as life span and length of the juvenile period, evolve. It is also about how behaviors can evolve. They change as we grow up— and so they change, to variable degrees, between species.

In writing this book I have been helped by many people who have supplied me with references, or have read parts or all of the manuscript, or have discussed areas in which they are experts and I am but a mere interloper. To these people I offer my sincere thanks: Carmelo Amalfi, Alex Baynes, Alex Bevan, Jenny Bevan, George Chaplin, Simon Conway Morris, Robert Craig, Brian Hall, Nina

Jablonski, Ron Johnstone, Mike Lee, Mance Lofgren, John Long, Mike McKinney, Dan McShea, Nancy Minugh-Purvis, Sue Parker, and Moya Smith. I would particularly like to thank Danielle Hendricks for producing all the fine drawings for this book (unless otherwise indicated). Robert Harington at the Johns Hopkins University Press was a constant source of encouragement, and I thank Carol Ehrlich and Celestia Ward for doing such a fine editing job. Last, but by no means least, my thanks for their forbearance go to my wife, Sue, and children, Jamie, Katie, and Tim.

So if you want a book about genetics in evolution, this is not the book for you. If you want a book extolling the virtues of natural selection, then this book may, in part, be for you. But if you want to know how evolution has produced the wing of a bat; the eye of a toad; the foot of an amphibian; the tiny, pathetic arms of *Tyrannosaurus rex*; or even your ability to read these words, then this may very well be the book for you. There is, after all, only one way to find out . . .

Shapes
of
Time

Prologue

IT WAS RAINING. YET ANOTHER SQUALL OF RAIN BLOWN IN FROM THE
Hebrides. For three hours we had tramped up and down heather-clad
slopes and across boggy valleys: a group of wet, bedraggled geology students from Aberdeen University, trekking across the Moine
Thrust in the west of Scotland. Here, hundreds of millions of years
ago, a vast slab of land had been heaved over tens of kilometers like a
gigantic rock sliding over a frozen lake. A slab of land which at that
time was attached to North America, but which later events had
ripped off and stuck to northern Europe.

   We were almost there. Almost back to the soggy warmth of our
bus, sitting enticingly just a hundred meters below us. Beyond rose
the dark outline of the mountain of Canisp. Closer lay a little loch:
Loch Awe. And past it snaked the road from Ullapool in the south,

heading north to the haven of Inchnadamph Hotel, with its blazing fires and a collection of malt whiskeys to make your eyes water. As the sun came out and the rain eased, we began to leap down the long slopes, the heather slapping our legs. Within minutes we had bounded through a few million years of time, to the younger Cambrian rocks deposited a little over 500 million years ago that lay beneath the Moine Thrust.

The road, at last. Where we came out there was a little quarry. We hadn't noticed it from on top of the hill. A quarry—that sounds rather grand. On the scale of this countryside, it was really little more than a tiny nick in the mountain. What were the rocks? "The Fucoid Beds," someone said. "Hey, there are meant to be fossil trilobites in those rocks." For a bunch of Scottish geologists who had cut their teeth on some of the ancient crystalline rocks of Europe, looking for fossils was deemed to be something only a soft Sassenach would do. OK, I was that Sassenach, so I had an excuse. "Come on, let's give it five minutes." OK. A crack on the rock with a muddy hammer. The mudstone splits open. Nothing. More hits. More nothing. People began to drift back to the bus. A last hit. And there it is. Staring out at the world after 540 million years trapped in rock—a trilobite.

Even though I had been collecting fossils since I was nine, there was nothing quite like finding my first trilobite. It must be something about the eyes, about the fact that you can look into the eyes of a beast who, when those eyes were working, saw a world more than half a billion years ago, so very different from the bleak Scottish landscape of today. Adding to the fascination was the realization that this was an olenellid trilobite, one of the earliest members of this extinct group of arthropods.

The hunt was on. More rocks cracked, and soon there was a good little collection. I went back many times to that quarry, usually deeper into the year, when the northern days were much longer. There was almost enough light that, given sufficient energy, you could collect until midnight. The fossils that the quarry yielded up were all molted heads: some large, others small. Looking at the different specimens was much like looking through a book of family photographs. The different-sized heads represented different stages in the trilobites' life history—juveniles of different ages, and adults— each fossil like an ancient snapshot. The challenge was, of course, as it would be in any pile of photos of a family that you didn't know and that had been taken over a long period of time, to sort out the ancestors and descendants, determining just who was related to whom.

As a paleontologist trying to unravel the evolutionary history of a group of fossils, I necessarily have to work in ways that other scientists often perceive as being unscientific. How, they argue, can you carry out the experiments that allow your hypothesis to be refuted? No problem: evolution and natural selection had long ago carried out the experiment—more than half a billion years ago, in this case. It was up to me to interpret the results.

Olenelloides armatus, an Early Cambrian paedomorphic trilobite from northwest Scotland.

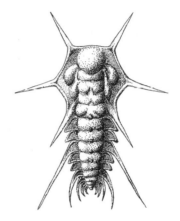

Four species seemed to be present in this collection. Three had been found elsewhere and named during the last century. One was a new species. Add to these a most bizarre little relative, called *Olenelloides,* which had been found not far away in the same rocks late in the last century, and there were five species in all. Looking at them one day I suddenly realized that placed in order, from the odd little spiky *Olenelloides* (which had been variously described as "primitive," "aberrant," "degenerate," or—and here was the clue—"larval" or "immature") to the "normal" type of *Olenellus,* this range of adult trilobites looked remarkably like the growth series of the "normal" *Olenellus.* In other words, some of these adults resembled juveniles of the "normal" type, at different stages of development. It was as if some had been frozen, Peter Pan–like, in time. Some, like *Olenelloides,* resembled very young juveniles; others were like adolescents. Yet each was much bigger than the corresponding growth stage.

From then on I was hooked. And it has been this relationship between evolution and an organism's developmental history, from conception through embryonic development, birth, and juvenile and adult development, that has fascinated me for the last twenty-five years. I soon came to realize that here was an area of evolutionary theory that had been very much neglected, at the expense of genetics and natural selection. Although popular in the nineteenth century, particularly through the work of the German scientist Ernst Haeckel, who promulgated his so-called biogenetic law, "ontogeny recapitulates phylogeny," few evolutionary biologists gave it much credence in the twentieth century. It was the Darwinian view of natural selection, followed by the rise of genetics in the 1930s, that consigned development to the evolutionary theory scrap heap. To most twentieth-century biologists, genetics and natural selection were the two fundamental aspects of evolution. But as a paleontologist who saw the end result of evolutionary processes that had operated over millions of years, I needed to know what it was that the genes were affecting. What was providing the raw material for natural selection to work on? Most analyses of evolution have focused almost exclusively on adult characteristics. Yet evolutionary pressures operate on an organism from the moment of its conception, right through to when it dies.

1

The Evolving Embryo

**Natural selection
acting on genes
is probably not
the whole story.
It's too simple.
Other forces
are also at work.**

Michael Crichton,
The Lost World

LIFE HAS EXISTED ON THIS PLANET FOR MORE THAN 3,500 MILLION
years. From the first single-celled bacteria a bewildering array of
other bacteria, plants, fungi, protists, and animals has evolved, from
trilobites to turnips, and from ammonites to azaleas, in a cascade of
increasing diversity. With the evolution of countless species of organisms came the colonization of a myriad of niches. But many of
these niches were built on shifting sands, and were forever changing
in their size and multidimensional shape. For any species a niche is
fashioned from the nature of the organism and that part of the environment with which it interacts. As the character of the environment
has shifted over countless millennia, so the sizes and shapes of the
animals and plants which inhabited them have likewise changed; and

as their morphologies have changed, so too their behaviors and their interactions with other species.

One overriding factor has been the arbiter in determining the composition of ever-changing animal, plant, bacterial, and fungal communities: time. And it is through the vast tracts of geological time that the diversity of organisms has increased, irrespective of whether new species evolved more complex structures than their ancestors, or whether they evolved more simplistic ones. Occasionally stupendous mass extinctions have punctuated life's seemingly endless ride through time. But life has accelerated into another burst of evolutionary fervor as it has hurtled on down time's roller coaster.

Yet as well as carving out whole communities of organisms, many now long extinct, time has been an equally potent force in evolution on a much smaller scale, in etching the quintessential character of each species at the level of the development of an individual organism. For the life of an individual of any species is, in many ways, a microcosm of the life of a species as a whole. Both individuals and species inherit many of their morphological traits from their ancestors. Each may give rise to descendants. And each suffers the ultimate, inescapable fate—extinction. Just how much the life history of individuals, from their conception to their death, is inextricably intertwined with the life history of species is the crux of this book. For it is an aspect of evolution that, although long recognized, has tended to have been pushed into the backwaters of evolutionary theory over the last century and a half.

Ontogeny: From Conception to Death

For those of us with a never-ending fascination with fossils, perhaps one of the greatest delights we can experience is to crack open a rock with a hammer and expose, if we are lucky, a fossil trilobite. Trilobites hold a particular fascination for many people, not only because of their antiquity, dating back to the dawn of animal life on this planet, but largely, I suspect, because many possess eyes. Although the caption to a beautiful photograph of one of the earliest (and largest) trilobites, *Paradoxides davidii*, in David Attenborough's "Life on Earth" would have it that such animals were blind, this was not the case. Even the very first trilobites, dating from rocks about 550 million years old, had well-developed eyes, set on raised plinths on either side of the head shield. Behind the head followed a multisegmented body and a shieldlike tail. The eyes possessed by these trilobites were, like those of other arthropods, such as insects, crabs, and lobsters, compound eyes. Each of the paired eyes comprised many separate lenses. And as the trilobite is released from its rocky tomb and the dust blows from these multifaceted eyes, it stares out of the rock from over hundreds of millions of years ago. This, after all, is the time scale on which great evolutionary changes, such as the appearance and disappearance of whole groups of animals, like trilobites, occurs.

At first sight, turnips would seem to bear very little in common with trilobites. Dig a turnip from a field. True, it does not look back at you with the hard, crystalline stare of the fossil trilobite, nor does it possess the trilobite's ancient pedigree. However, time also plays just as crucial a part in the appearance of the turnip as it does in the trilobite, and, indeed, in its innate attractiveness. As well as carrying within its hard, pungent flesh the genetic imprint of its evolutionary history, each and every turnip, like every other organism that has existed on this planet, carries time's imprint, which is also expressed in a short developmental history. With a freshly dug-up turnip we are looking at the culmination of its *ontogeny*—in other words, the end result of the growth and development of that particular turnip from a single cell to the dubiously palatable organism it has become as an adult. Like all other species, both the trilobite and the turnip carry within their genes their own evolutionary histories—and both, as individuals (one long dead, the other, if you are that way inclined, soon to be eaten), their individual life histories. Thus time links the individuals' life histories and their species' evolutionary history.

Time, in evolution, really is of the essence. This may seem to be self-evident, yet in many respects the role of time in evolution has been rather surprisingly neglected, particularly in terms of individual life histories. While in the biological sense of evolution we mean changes over time in the nature of the organism, whether it be morphological, physiological, or behavioral, many contemporary studies have focused on the temporal changes between species or between higher taxa. For example, we can talk about a snail species *x* that lived 2.5 to 1.5 million years ago and which evolved into snail species *y*, which existed until half a million years ago. Often overlooked in studies of evolution is what actually underlies the changes in shape and size between species and also the part played by changes in the timing and rate of development of individuals, from their fertilization, through the juvenile and adult phases to their deaths, and how such changes are a significant part of the evolutionary process.

Evolution and Natural Selection

In terms of our current understanding of the nature of the Universe, can there really be anything quite as amazing as the fact that life, as exemplified by the trilobite and the turnip, ever became established on this little rocky outpost at the far side of just one of the countless galaxies spinning through the Universe? For all our sophisticated technology and countless man-years of listening to the echoes from the far-flung reaches of the Universe, there is still no indication that we are anything other than alone in this expanding void. As Democritus wrote circa 420 B.C.:

> Apparently there is colour, apparently sweetness, apparently bitterness; actually there are only atoms and the void.

Such thoughts have the tendency to send a collective shiver of panic down our spines and to provoke a most unpleasant case of cosmic vertigo. Perhaps this is why many scientists, rather than peering back through time by looking at ancient light emissions from stars a dinosaur's evolutionary history away, are happier examining time more introspectively through the evolution of life on Earth. The ultimate result of such introspection, while sitting here on this spinning lump of rock, water, and gas as it hurtles through space, is the age-old pondering of not only what we are doing here, but, more pertinent to this book, where we have come from. Such questioning seems to have clogged up our collective consciousness ever since we as a species could think beyond what we were going to eat for our next meal, and how we were going to avoid becoming something else's next meal.

These days, when the concept of evolution enters people's consciousness, I suspect that quite often what appears in their mind is not the thought of some insignificant little cluster of cells mutating into another, slightly different cluster of little cells, but a much more complex cluster of cells in the form of a somewhat austere gentleman with receding hairline and long, white beard—one Charles Darwin.

To most of us Darwin epitomizes evolution. Any time B.D. (Before Darwin) is considered by most people to have been the Evolutionary Dark Ages, when there was little, if any, coherent view of evolution. Most pre-Darwinian biologists, it is generally inferred, were creationists, or too interested in the minutiae of biology even to contemplate that one species could ever have evolved from another. Or so it seems. But with a closer examination of the writings of some of the leading anatomists, embryologists, and naturalists of the early nineteenth century it becomes apparent that the question of the relationships between species was as fascinating to them as it was to Darwin and his contemporaries. Furthermore, the concept of evolution certainly did exist before Darwin's time. The trouble was that in many respects the pre-Darwinian biologists had been going round and round in circles for decades and were trapped by the limits of their theories, as I discuss in the next chapter. But then in the middle of the century Charles Darwin blew in with a fresh and novel approach. The result was that most free-thinking biologists seized onto Darwin's new ideas with all the fervor of a hungry kitten latching onto a piece of raw meat, and many of the earlier attempts to interpret the

Cartoon from *Punch* magazine, 1881

MAN·IS·BVT·A·WORM·

interrelationships between species were cast on the scrap heap of evolutionary theory.

What Darwin argued seems in many ways, with the comfortable benefit of hindsight, to be very obvious: that competition between individuals and "survival of the fittest" in terms of the individual's adaptation to the environment through natural selection were the overwhelmingly dominant factors in directing the course of evolution. This in many ways contradicted what other biologists had been arguing in the early part of the nineteenth century: that the key to relationships (and thus, *ipso facto*, evolution) lay in how the morphological changes that occurred between organisms took place, and in particular how changes in the rate and timing of embryological growth of different species could indicate interrelationships between species. Here, in some way, was the dichotomy of evolution: on the one hand, what was happening *intrinsically* to the organism—the internal forces directing changes in the course of morphological and physiological development; and on the other hand, the *extrinsic* factors—those external forces, such as competition and predation, resulting in natural selection. It was the latter forces that Darwin saw as molding and shaping each subsequent generation. To Darwin, natural selection implied "only the preservation of such variations as arise and are beneficial to the being under its conditions of life" (*Origin of Species*, 6th ed., p. 63). In those times before the discovery of the existence and importance of genes, the internal factors that were responsible for the origin of such variations were hardly considered by Darwin. And as I point out in the next chapter, only about 2 percent of his book was devoted to the role that embryological development played in evolution.

Interestingly, although we equate Charles Darwin with the term *evolution*, it was a word he rarely used. In the first edition of his *Origin of Species*, published in 1859, he spoke of "descent with modification" to describe the changes from one to species to another, rather than evolution. In fact, surprisingly, the term appeared only as the very last word in his book: "There is a grandeur in this view of life, with its several powers, having been originally breathed by the Creator into a few forms or into one; and that, whilst this planet has gone cycling on according to the fixed laws of gravity, from so simple a beginning endless forms most beautiful and most wonderful have been, and are being evolved."

However, by the publication of the sixth edition in 1872 he used the word "evolution" twice, reflecting, as I discuss shortly, the increasing use of the word during the latter part of the nineteenth century to describe Darwin's descent with modification.

While the original, prebiological meaning of the word portrays the act of unrolling or unfolding (from the Latin *evolutio*, an unrolling), its use in biology has come to be associated with two different concepts: change and time. As Stephen J. Gould of Harvard University has pointed out in his book *Ontogeny and Phylogeny*, in its biological sense the word was first used in 1744 by the Swiss botanist, physiologist, poet, and lawyer Albrecht von Haller (1708–1777).[1] In his *Hermanni Boerhauve praelectiones academicae*, Haller wrote:

> But the theory of evolution proposed by Swammerdam and Malpighi prevails almost everywhere. . . . Most of these men teach that there is in fact included in the egg a germ or perfect little human machine. . . . And not a few of them say that all human bodies were created fully formed and folded up in the ovary of Eve and that these bodies are gradually distended by alimentary humor until they grow to the form and size of animals.

Haller's choice of the term "evolution" derived, quite appropriately, from the Latin term meaning unrolling, or unfolding. As Gould notes in *Ontogeny and Phylogeny*, the transformation of the word from its original biological meaning coined by Haller is in itself a fascinating story. For to Haller and his contemporaries, like Charles Bonnet and Joseph T. Needham, it was used, as Haller's quote shows, for a simple description of *embryological* development (what we would now call *ontogeny*). This is quite a different meaning from its currently accepted usage in describing interpopulation or interspecific change.

In this original embryological context it was used in 1791 by Charles Darwin's grandfather Erasmus to describe "the gradual evolution of the young animal or plant from the seed." However, during the 1820s and 1830s there arose some confusion over its use in an embryological sense. Embryologists at this time argued that Haller didn't really believe that the human embryo was an absolute miniature of the adult form, rather that there was some structural change during development; therefore, Haller's use of the word implied some degree of change in shape, not just size. But this change was during the development of an individual, not the transformation of one species into another.

The person generally considered as having been responsible for changing the use of the term was Herbert Spencer (1820–1903) in the mid-nineteenth century. Although trained as an engineer, Spencer was a highly influential writer, social philosopher, and journalist.[2] He was using the term "evolution" in its current biological sense in 1852 in his essay "The Development Hypothesis," stressing the increase in complexity and interaction with external factors, rather than the purely internal view of "transmutation" espoused by other writers. In his autobiography, published in 1904, Spencer records that the early-nineteenth-century embryologist Karl Ernst von Baer (see Chap. 2) made him aware that "the law which holds of the ascending stages of each individual organism is also the law which holds of the ascending grades of organisms of all kinds."

Spencer therefore expanded the use of the term "evolution," from a term which previously had been restricted to explaining the changes that occur during the development of an individual, to the changes that occur during the transformation of one species into another. This, somewhat paradoxically, was in itself a recognition, as had been argued so forceably by many biologists before Darwin, that the development of an organism and the "transformations" that occurred between species, were inextricably linked. Yet the term that had been used just once by Darwin has come to be regarded as being largely synonymous with his concept of natural selection—the extrinsic aspect of evolution, rather than its original intrinsic meaning.

In "The Development Hypothesis," as well as in his later writings, Spencer, when writing about evolution, stressed the concept of change from an "indefinite, incoherent homogeneity" to a "definite, coherent heterogeneity." He came, therefore, to associate the term with "progressive" change, a view still held to a large degree by most laypeople. In many ways, Spencer's "progressive" view of evolution is a natural consequence of the extrapolation of the embryological context of the word, as used in the early nineteenth century. As the importance of embryological development declined, so too the meaning of the word "evolution" changed, from progressive to adaptationist. Spencer's influence was such that thereafter a range of writers, from Lyell to Wallace and Huxley, and even Darwin himself, equated the term "evolution" with "transmutation." With the rise of the study of Mendelian genetics in the twentieth century, usage of the term was extended even further to encompass genetic changes that occurred within populations.[3]

This slight etymological digression into the derivation and subsequent change in usage of the word "evolution" is more than just a self-indulgent clarification of some arcane semantic question. It encapsulates the tremendous change in emphasis that has taken place in evolutionary studies from the early nineteenth century to the present day, from the inward-looking concentration on the relationship between embryology and species relationships to a dominance in the external role of natural selection.

Evolution from Within

Most other books on evolution, and practically all so-called popular books on the subject, have looked at evolution either purely from this extrinsic point of view or combined with the role of genetics. But this book aims to reset the balance, to some degree, by examining evolution from the inside: by looking at just how changes in size and shape of animals and plants occur; what factors other than natural selection operate; and what factors influence these changes. But do not be misled: this is not a book about genetics. While the nature of our genome will be the ultimate factor in affecting how we grow and develop and so what we come to look like as adults (I discuss this in Chap. 3), this too, is still only part of the evolutionary equation. Since genetics rose to pre-eminence in the 1930s, its interplay with natural selection has dominated evolutionary theory. The middle ground, of how organisms actually change their shapes and sizes, has largely been neglected.

The so-called neo-Darwinism, or "modern synthesis," sees the most important aspects of evolution as being *mutation* (the spontaneous origin of new genetic variants, due to changes in the genome), *natural selection, gene flow,* and *genetic drift.* The latter two concepts involve the introduction of new genes into a population and the random changes in the genetic composition of a population. What I argue in this book is that in this modern synthesis a critical factor in the equation is left out: a missing link in evolutionary studies, the role that changes in the patterns of development of an organism play in evolution. In other words, how do variations in the genetic makeup of species influence development and produce the changes in shape and size of animals and plants to allow them to be susceptible to natural selection? This missing link is the third, and central, factor in the triumvirate of evolution: genetic changes in the timing and rate of development producing variation that is acted upon by natural selection.

Let's consider a culinary analogy. Consider the three basic elements of evolution: genes, the products of changes in organisms' developmental patterns, and natural selection. Then imagine that the genes are chefs in a restaurant; the products of the developmental changes are analogous to the meal that the chefs have created; and the diners at the restaurant play the part of natural selection. These gourmands vote with their taste buds and decide whether or not a particular meal will be produced again, in much the same way that natural selection determines whether an organism is "fit" enough to survive

The advertising industry's view of hominid evolution—adults only. Reprinted by permission.

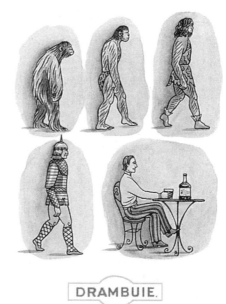

and to pass its genes on to another generation. Similarly, if the customers like the meal, then the chefs will make it again. If they don't, then the recipe will be doomed to extinction. Likewise, if natural selection doesn't "like" a particular morphotype, it will not be preferentially selected and will be thrown on the evolutionary scrap heap. Now, for far too long evolutionary biologists have concentrated their attention on genes and on natural selection. It's as if a food critic were concerned simply with the characteristics of the chef and the peculiarities of the customers, and took virtually no notice of the food and how it was produced. Yet without the food there would be no chef and no customers. All three need each other to operate as a fully functioning, productive entity. So it is with biological evolution.

When it comes to perceptions as to how living things have evolved, what do most people think of? My bet would be that the vast majority would visualize a picture, based for the most part on some advertising agency's view of evolution, of man evolving from the apes. Just picture any of the illustrations that have been used to chart the stately progression of *Australopithecus* to early species of *Homo*, through to *Homo sapiens*, finally through to modern man sitting in front of a glass of Drambuie, or whatever. Inevitably such illustrations depict adults (almost invariably males). This reinforces a view which is espoused in so many textbooks and scientific papers on evolution—that the adult male form of one species evolves into the adult male form of another. The embryonic phase and the subsequent juvenile, preadult phase seem to be viewed almost as unfortunate, rather embarrassing states that the organism has to pass through in order to achieve the state of higher being (i.e., adulthood). Presumably one is expected to believe that only adult males evolve—not juveniles or females.

The other aspect of evolution that probably enters people's minds when they think about evolution is genetics. All morphological change, after all, is under the control of genes: those slightly mysterious bits in our cells that we can never hope to see, but which somehow affect what we look like and, to a certain degree, how we act. So we usually think that the evolution of our adult male ape into an adult male *Homo sapiens* occurred by an alteration in some of these genes: a genetic mutation. A change in the genes or how they act in some way occurs, and this produces a creature of a different shape and perhaps a different size. If this new form is better suited to its environment than its predecessor or other coexistent forms, and the genetic change can be passed on to future generations, then a new species evolves. This is what is known as the "modern synthesis"—genetics + natural selection = evolution. That's fine, but there is one slight gap in this piece of logic. How do we get from the genetic change to the morphological change? How is this grand transformation in shape and size achieved—how did I end up with a larger body size and bigger brain than *Australopithecus?* What was the mechanism between the genetic change and the natural selection that resulted in me, sitting at a desk, typing on a keyboard while listening to motets by Bruch? How do changes in the genetic makeup of one species cause changes in descendant adults, leading ultimately to the evolution of a new species? And what part is played in this drama by time, acting on the timing and rate of growth of different parts of my body, from the time of conception until the time that maturity is reached and morphological development virtually stops?

As we know, to Charles Darwin the driving force behind evolution was natural selection—the nature of the environment in which the organism lives and with which it interacts, determining whether one individual is more successful than another. And natural selection has, quite rightly, remained one of the cornerstones of evolutionary theory. But what causes the slight changes between individuals that give one a slight edge over its neighbor? How does this so-called intraspecific variation arise? Genetic mutations may well play a part, but what is the nature of these mutations and how are they translated into changes in the appearance of animals or plants?

In our currently accepted biological notion, evolution means change. We think of it as the evolution of one species from another, or of one population of individuals from another. For example, such evolution may result in one species of fly evolving from another by a subtle change in the size and shape of its wings, and the size and shape of its visual surface. Slight changes, maybe, but perhaps enough to result in ecological separation of the species and different behavioral patterns. But if we focus on the individual of an organism—focus in on yourself, if you like—then each organism undergoes a phenomenal degree of change from its moment of conception until it dies. More often than not the morphological and associated behavioral changes that blossom in an individual organism during its development make the differences that prevail between the adults of related species pale into insignificance. While there are the dramatic changes that occur between organisms that metamorphose, such as the very obvious differences between a caterpillar and a butterfly, or between a tadpole and a frog, even in mammals, where the change is less abrupt, it can be appreciable. You might think that there has been a dramatic change in appearance from an "ape man," such as *Australopithecus*, to modern man. But this is nothing when compared with the morphological changes that occur as an individual of a single species grows up.

To test this idea, go look at yourself in the mirror. Not a pretty sight, perhaps, but try to let your imagination take you back to the time before you were born. Go back to the first few weeks of your existence following conception. Think about what you would have looked like thirty days after conception. You were obviously much smaller than you are now, about the size of a pea; your brain had scarcely started to develop; you would have had three pairs of arterial arches in your neck region, which fortunately for you disappeared. If they hadn't, you would now be the proud owner of a set of gills. You would have had a well-developed tail, and your arms and legs would have been little more than tiny buds. Through the next twelve weeks your head would have increased greatly in size, while your limbs would have undergone an enormous relative increase in length and sprouted toes and fingers. By the end of these next twelve weeks, your face would have become recognizably human. We rarely think of these remarkable changes that we have undergone to get us to what we have become today. This is our *ontogeny*—our growth and development from embryo through juvenile to adult.

Within certain constraints individuals of every species, our own included, undergo similar relative changes in body parts to other individuals throughout ontogeny. But just consider for a moment what would have happened if the orderly rate and timing of development of your body parts had been altered during your development. If the rate at which a feature that changed mostly during the latter part of your juvenile development had been slightly altered you might have ended up with a slightly longer leg, or perhaps a shorter arm, than usual. However, if a critical feature that formed early in your embryonic development had perhaps started its growth either a little earlier or a little later than normal, or had grown at a slightly different rate, then the consequences for your adult shape would have been tremendous. If you had persisted as a viable individual organism, you might look quite different from other members of your species. Yet genetically you would be virtually identical. The point that I am trying to make is that slight changes in growth rates or timing at critical periods of development can, with very little genetic change, produce profound *phenotypic* effects (in other words, greatly affect what the organism looks like). The result is that a descendant can look potentially quite different from its immediate ancestor, while having suffered only minimal genetic change. After all, as it has been pointed out on many occasions, genetically humans and chimps are almost 99 percent identical.

This concept of changes in the timing and rate of development is a fundamental, but these days a much neglected, aspect of evolution. However, if, as I argue, it is so important, why has it been overlooked for so long? There are many reasons for this, mostly of a historical nature. In part it may well lie, particularly with nonbiologists (as well as a fair smattering of biologists, to whom the whole idea is quite foreign), in what it is called. While the theory for the origin of the Universe is beautifully described as the "big bang" theory; while "natural selection" resonates with elegance and clarity; and while the terminology in physics derived for subatomic particles, such as quarks, has (dare I say it) a charm all of its own, the relationship between organisms' ontogenies and their evolutionary history is manacled with a less than euphonious term: *heterochrony*. Literally meaning "changing time," it is hardly the sort of term which on hearing it is likely to send anyone into paroxysms of delight. But get used to it. It is a word that appears many times throughout this book, so that when you reach the final page the word "heterochrony" will fall as readily off your tongue as water falling down a drain.

The concept of heterochrony was first introduced by one of the major protagonists in the study of the relationship between development and evolution, the leading late-nineteenth-century German anatomist Ernst Haeckel (1834–1919). Haeckel defined heterochrony essentially as a displacement in time, or dislocation of the phylogenetic order of succession.[4] However, as with the biological meaning of the word "evolution," so "heterochrony" has been used with a variety of meanings (see Chap. 2). In recent years, though, particularly following the renaissance in interest in the phenomenon engendered by Harvard paleontologist Stephen Gould in *Ontogeny and Phylogeny*, "heterochrony" has been used in a very general sense to describe the displacement in time of the appearance of a particular feature relative to the time that the same feature appeared in an ancestral form.

Woodcut of painting by Franz von Leubach, 1899, of the father of the biogenetic law, Ernst Haeckel

In its simplest form we can view the "amount" of change that an individual goes through as it grows up as being either more or less than its ancestors'. If it grows at a faster rate or for a longer time, it might produce more cells. On the contrary, if the "amount" of growth it undergoes is reduced, by growing more slowly or for a shorter period of time, it may produce fewer cells. In terms of the animal's or plant's morphology (its shape and size), or just what it looks like, the species that has undergone more growth could be regarded as being more "complex," whereas the species that has undergone less growth would be more "simple." The concept of complexity in evolution has cropped up time and time again, and has rarely been very well resolved. The evolutionary success of a group is often considered to be reflected in an increase in the morphological and physiological complexities of the sensory systems. Larger size and greater morphological complexity of the brain are also viewed as reflecting evolutionary "success." While such a simplistic view of bigger being more complex and smaller less complex has tended to be widely held, in terms of evolution this is exceedingly misleading. As I shall show, the view that morphological simplicity equals evolutionary simplicity is just wrong. The smallest, morphologically most simple species can be the most specialized and ecologically complex of species. Thus, the age-old view of evolution meaning getting bigger and better is exploded when the role of changing ontogenies is taken into account. For evolution can just as well be toward the smaller and simpler as toward the larger and more complex.

Viewed on the grand scale of evolution from deep time, billions of years ago, through to the myriad of species that have occupied this planet during the last few hundred million years, evolution has progressed from the simple to the more complex. It has done this firstly by increases in cell complexity,[5] from the bacterial prokaryotic cells that are little more than bags of DNA to eukaryotic cells (cells containing complex structures within their framework, such as the nucleus and the mitochondrial powerhouse, which have larger genomes). Then within eukaryotes, genome size has increased, being greater in animals and plants than in protists. Greater complexity is further demonstrated by the evolution of multicellularity and by an increase in the number of cells. With this has come an increase in the diversity of cell types, reflected in an increase in anatomical, physiological, and behavioral complexity, through simple, jawless fishes to amphibians, reptiles, birds, and mammals. But when little segments of the grand picture of evolution are viewed through a paleontological magnifying glass, things are not quite so clear. On a bacteria-to-bat comparison, the bat wins hands down in the complexity stakes. But is anatomical complexity the key to evolutionary success? Bats have existed for 50 million years; bacteria for at least 3,500 million. And there are many places on Earth that bacteria have colonized to which bats could never hope to go.

Even though there has been a long tradition of proudly viewing our own species as the pinnacle of evolution—what with our most sophisticated and complex of brains and behavioral patterns—it would, as one of the great evolutionary biologists of the twentieth century, George Gaylord Simpson, has said, "be a brave anatomist who would attempt to prove that Recent man is more complicated than a Devonian ostracoderm."

Even defining complexity poses many problems. Is a bat more complex than a bacterium? To be sure, it has more cells, but it also has more different types of cells, so that could be, as John Bonner of

Princeton University has suggested, one criterion to use. More cell types means more specialized functions and increased behavioral complexity. On average, larger organisms have more types of cells than smaller ones. So, then, does large body size mean more complexity? And what produces larger body size? When the fossil record has been used to assess whether or not organisms become more complex through time, the results have been very equivocal. Dan McShea, of the Santa Fe Institute in New Mexico, analyzed this by looking at changes in morphological complexity of the vertebral columns of a range of animals—camels, whales, squirrels, pangolins, and chevrotains—and found that over the past thirty million years there have been no trends toward increased complexity.[6]

Similar studies carried out by George Boyajian, of the University of Pennsylvania, and Tim Lutz, of West Chester University, Pennsylvania, on the frilly suture lines in ammonoid shells (which reflect the degree of complexity of the walls that separate chambers within the shell) show trends toward both increases and decreases in complexity.[7] There has been a long history of interpreting such changes in ammonoids in terms of heterochrony. While it was the fashion in the nineteenth century to see biological evolution as a never-ending increase in complexity, support for this notion was found in the trends toward increased development of suture lines in these extinct, shelled relatives of octopuses and squid. In the twentieth century, when the opposite was argued, that reduced "amount" of growth was the key to evolutionary novelties, ammonoid sutures were again trotted out to demonstrate this point. But as Boyajian and Lutz show, increases *and* decreases in sutural complexity occur, reflecting selection for whatever is appropriate at that place and at that time. And we can only hope to explain these changes in complexity over the vastness of geological time by understanding the nature of organisms' ontogenies—individuals' biological time. For only by so doing can we hope to explain the underlying mechanisms that control changes in morphological complexity over time.

The idea of heterochrony has had a long and checkered history, upon which I shall elaborate in the next chapter. Is it the crucial link between the functioning of genes and the production of the raw material upon which natural selection can operate? Can it help us interpret changes in complexity? It was a favored notion, in a particularly rigid form, of many anatomists in the last century. However, with the rise of "Darwinism" and its influence on promoting external rather than internal factors in evolution, the influence of their ideas on evolution diminished markedly. And with their decreased influence came a general lack of interest in trying to appreciate exactly how changes in ontogenies over time have played such a significant part in evolution. However, over the last decade many biologists and paleontologists have come to realize that once again it is to the developmental history of individual organisms that we must turn if we are to understand fully how animals and plants evolve. For it is not just the adults of a species that evolve but the entire growth sequence. Indeed, evolution could well be defined as "a sequence of ontogenies that change through time."

Here, I believe, lies the secret of how life has diversified to such an astounding extent on this planet over the last three and half billion years, particularly within the last 550 million years. For it is such changes in the timing and rate of development of parts of the body that have provided the raw material for natural selection and allowed this panoply of life to evolve into such a wide range of environments and habitats.

2

The Topsy-Turvy World of Dr. Haeckel and Dr. Garstang

In the youth of every theory everything is beautifully clear, and ideally simple. As time goes on we are compelled to drop one idea after another, until it almost seems that the whole will be lost.

J. P. Smith, "Acceleration of Development in Fossil Cephalopoda"

ANY ATTEMPT TO WRITE ABOUT THE RELATIONSHIP BETWEEN ORGAN-isms' developmental histories and their evolutionary histories will forever be constrained by the tyranny of the past. It is no easy thing for evolutionary biologists today to exorcise the specters of some of the great nineteenth- and early-twentieth-century anatomists and embryologists who peer over their shoulders. Nor should they be eager to do so. For while, with the benefit of hindsight, it might be argued that we have been led up the evolutionary garden path on more than one occasion, the part played by these estimable scientists in the development of ideas on heterochrony should not be underestimated. To understand why heterochrony has been given such a small bit part in "the modern synthesis," we must explore the vacillating fashions in evolutionary theory over the last two centuries. For the study of evolution is as much constrained by the vagaries of the past as is the evolution of life itself.

Haeckel and the Biogenetic Law

The central figure whose specter certainly casts the greatest shadow is Ernst Haeckel. A leading biologist in the mid-nineteenth century (he even thought up the word *biology*), his most influential tome, *Generelle Morphologie der Organismen*, was published in 1866. Within this book was written one short phrase which is pivotal to studies of heterochrony. On the one hand, this phrase can be construed as being the quintessential articulation of the culmination of half a century's studies on the relationships between organisms' embryological development; on the other, it had a disastrous effect on studies of the subject for the ensuing one hundred years. What Haeckel wrote was that "Die Ontogenesis ist die kurze und schelle Rekapitulation der Phylogenesis, bedingt durch die physiologischen Funktionen der Vererbung (Fortpflanzung) und Anpassung (Ernährung)." Translated, this reads: "Ontogeny is the brief and rapid recapitulation of phylogeny, dependent on the physiological functions of heredity (reproduction) and adaptation (nutrition)." Remember, for *ontogeny* read *development*; for *phylogeny* read *evolution*.

Writing just seven years after the publication of the *Origin of Species*, Haeckel had attempted to synthesize the basic tenets of many of the early-nineteenth-century embryologists within the framework of Darwin's concept of the importance of adaptation in evolution. Underlying the first part of his statement was the concept, stressed so forcefully by Karl Ernst von Baer in the 1820s, that the earlier the embryological stage of an organism, the closer the similarities between members of disparate groups. Yet what Haeckel's statement boiled down to was the abbreviated dictum that *ontogeny recapitulates phylogeny* (development recapitulates evolution). In other words, early embryos of mammals, for instance, would be viewed as passing through developmental stages that resemble, at one time, worms; at another, fishes; at another, reptiles, before passing into a "mammalian" stage. Such a concept reinforced the Spencerian view of evolution as one of "progress" and increasing complexity. This maxim became known as the "biogenetic law." Such was its influence in biology that it continued to be taught through much of the twentieth century long after it had been discredited as an all-embracing law. Indeed, I know from my own personal experience that the concept that ontogeny recapitulates phylogeny was being taught in at least one university in the U.K. in the early 1970s. As we shall see, not only did this strait-jacketed approach, which saw all change as being a progressive addition of morphological complexity of the entire organism, severely constrain future studies, but Haeckel's original concept of the dual roles of the internal processes (developmental change) and external processes (adaptation) was ignored.

But in the same way that evolution existed long before Darwin, so the concept of recapitulation existed long before Haeckel. The intimate relationship between an organism's developmental and evolutionary histories had been viewed in a rather myopic fashion from a single perspective throughout the nineteenth century. The commonly held view at the time was that complexity progressively increased from "primitive" life forms, to the lower classes, thence through the upper echelons of the animal kingdom. It goes without saying that mankind was viewed as the pinnacle of Creation, being a reflection of God in all His glory. The embryological development of "higher" animals was thought to encapsulate all the adult stages of life forms that were perceived to be lower on the scale of organismic complexity, and which had evolved earlier. From this perspective it was thought that during its growth and development the human embryo passed through, first, a fish stage, followed by a reptilian stage, before finally attaining its "higher" mammalian state. This, in essence, was the concept of recapitulation that, in various guises, formed the foundation of developmental biology throughout much of the nineteenth century.

Such a concept did not spring fully formed into the consciousness of anatomists and naturalists at this time. In fact, it has a most illustrious pedigree, being traceable as far back as the time of Aristotle, for the concept of increasing complexity in the biological world was a central part of Ancient Greek philosophy. The Greek astronomer and geographer Anaximander (c. 611–547 B.C.), for instance, in his treatise on nature and growth, perceived the history of development of the human embryo as paralleling the appearance of the modern

human form, arising from hypothetical ancestors that were first encased in horny capsules that floated and fed in the water, before burst-
ing forth from their "fish-men" state to emerge and move onto land
as fully fledged humans.[1] Here Anaximander was in some ways presaging the nineteeth-century view of a comparison between the embryo in a fishlike stage, floating in its watery environment, and that
"level" of anatomical development in fishes. As Stephen Gould has
pointed out in his book *Ontogeny and Phylogeny*, Aristotle expressed
the idea of recapitulation as the sequence of increasingly more complex souls—which he thought of as "nutritive," "sensitive," then "rational"—that entered the human embryo during successive stages of
development. Aristotle compared the "nutritive" soul with plants, the
"sensitive" with animals, and the "rational" with humans. Although
in this regard Aristotle has been regarded by some people as the
great-great-grandfather of recapitulation, this is questionable. It is
debatable whether much can really be gained from trying to marry
such philosophical ideas with concepts that were generated from embryological studies carried out more than two thousand years down
the track.

The Early Recapitulationists

It has been suggested that the modern concept of recapitulation arose during the early nineteenth century
as "an inescapable consequence of a particular biological philosophy."[2] This was the school of *Naturphilosophie*, which existed in Germany at that time.
The followers of this philosophy possessed an uncom-
promising belief in "developmentalism"—a unidirectional flux that
moved from lower to higher, from initial chaos and thence to man. It
was therefore hardly surprising that philosophers of this school em-
braced recapitulation with open arms. The first notable proponent of
recapitulation of this school was the anatomist and embryologist
Lorenz Oken. Best known for his work on the embryology of pigs
and dogs, Oken proposed a classification of animals that was based on
a progressive addition of organs during development. He believed
that, after starting from nothing, there was an increase in the number
of organs, combined with a general increase in complexity. Oken believed that this occurred by a simple successive addition of organs in a
predetermined sequence. Under this scheme man was seen as the

summit of Nature's achievement, reflecting a microcosm of the whole world. In his *Elements of Natural Philosophy* Oken wrote:[3]

> During its development the animal passes through all stages of the animal kingdom. The foetus is a representation of all animal classes in time. At first it is a simple vesicle, stomach, or vitellus, as in the Infusoria. Then the vesicle is doubled through the albumen and shell, and obtains an intestine, as in the Corals.
>
> It obtains a vascular system in the vitelline vessels, or absorbents, as in the Acalephae.
>
> With the blood-system, liver, and ovarium, the embryo enters the class of bivalved Mollusca.
>
> With the muscular heart, the testicle, and the penis, into the class of Snails.
>
> With the venous and arteriose hearts, and the urinary apparatus, into the class of Cephalopods or Cuttle-fish.
>
> With the absorption of the integument, into the class of Worms.
>
> With the formation of branchial fissures, into the class of Crustacea.
>
> With the germination or budding forth of limbs, into the class of Insects.
>
> With the appearance of the osseus system, into the class of Fishes.
>
> With the evolution of muscles, into the class of Reptiles.
>
> With the ingress of respiration through the lungs, into the class of Birds. The foetus, when born, is actually like them, edentulous.

Another recapitulationist was the German embryologist J. F. Meckel (1781–1833). Arguably the most influential recapitulationist of the Naturphilosoph school, he believed that a single developmental pattern governed nature. However, he differed from Oken in not relating it purely to the addition of new organs. He viewed it more as an increase in specialization and coordination of the entire organism. Writing in 1821, in his *System der vergleichenden*, Meckel stated: "The development of the individual organism obeys the same laws as the development of the whole animal series; that is to say, the higher animal, in its gradual development, essentially passes through the permanent organic stages that lie below it."

To Meckel and his colleagues it was clear: the order that vertebrates appeared in the fossil record (fish-reptile-mammal), combined with the appearance of vertebrates after invertebrates, paralleled embryonic development. A strong belief in recapitulation was also prevalent among members of the French school of transcendental morphology. One notable member of this group was the medical anatomist Etienne Serres, who studied the comparative anatomy of the vertebrate brain. Having studied the nervous systems of vertebrates and invertebrates, he concluded that "lower" animals were the permanent embryos of "higher" animals. Like the German Naturphilosophs, Serres saw in embryonic development a "step by step march to perfection." However, Serres saw some specific exceptions to this overriding view of life in the form of "developmental arrests." One example of this was the state of undescended testicles in some men. Serres saw in this a parallel with the permanent state in fishes, where the testicles are retained within the body cavity. Likewise, malformations of the heart he saw as being comparable with the stages achieved by "lower" animals.

But this all-embracing recapitulatory view was not universally accepted. Among its most vehement opponents was the great nineteenth-century embryologist Karl Ernst von Baer who, while recognizing that similarities between animals' early developmental stages indicated some sort of relationship between such forms, argued that it was invalid to compare directly the adult form of one animal with the juvenile stage of another. Rather, he argued that "the embryo of a higher animal is never like the adult of a lower animal, but only like its embryo."

In his monumental tome *Entwickelungsgeschichte der Thiere*, which was published in 1828, von Baer set out to demolish many of the basic tenets of recapitulation. The principal argument that von Baer used was that the theory of recapitulation carried within it a fundamental error. He argued that arrested embryonic stages do not compare directly with the adult state of "lower" organisms, as would be expected from the theory. Von Baer showed that many embryonic features are specializations. Moreover, although there may be a few similarities between an embryonic structure of one organism and the adult of another, many other structures will not be comparable. An even more potent argument is that many features that occur in adults of "higher" organisms can actually occur in embryos of so-called lower organisms. Von Baer argued that at all stages of development, embryonic vertebrates are undeveloped vertebrates and do *not* represent any "lower" adult form. Perhaps the most perceptive argument that von Baer came up with was that embryological development is not just a simple case, as Stephen Gould puts it, of "a climb up the ladder of perfection"; rather, it is a differentiation from more generalized characters to more complex ones. The significant point to emerge from this is that the more general features shared between many members of a large group of animals appear *earlier* in development. By contrast, the more specialized characters appear *later* in development. So you cannot say, as the recapitulationists did, that the embryo of a "higher" animal resembles the adult of a "lower" animal, but only that it resembles its embryo.

Needless to say, most embryologists, anatomists, and paleontologists ignored von Baer's perceptive insights. Paleontologists soon latched onto the theory of recapitulation and tried to demonstrate its existence in every sequence of fossils they could lay their hands on. Among the first to do so was the great nineteenth-century Swiss paleontologist Louis Agassiz (1807–1873), having been influenced by lectures given by Oken. In one of the earliest and most important publications on fossil fishes, his *Recherches sur les poissons fossiles*, which was published between 1833 and 1843, Agassiz argued that the form of the tail in adult fishes, as observed in the fossil record, paralleled the development of the tail seen in modern fishes, from early embryonic stages to the adult stage. Thus, small juvenile teleost fishes have a simple tail; later juveniles a so-called heterocercal tail, where the upper part is larger than the lower; leading to a homocercal tail, where the top and bottom fins are the same size. This paralleled a perceived evolutionary sequence that Agassiz argued could be shown from the fossil record.

Agassiz' talents were not confined to studying fossil fishes. He also carried out pioneer work on echinoids (sea urchins) during the late 1830s and early 1840s. He observed how adults of "irregular" echinoids (such as sand dollars and heart urchins) have a mass of fine, hairlike spines. As juveniles, however, they have few, relatively large spines. In this regard they closely resemble the adults of cidaroid echinoids (slate pencil urchins), who possess just a few thick, long spines. These cidaroids, Agassiz noted, predated urchins like sand dollars in the fossil record.

The Heyday of Recapitulation

What really separated the early-nineteenth-century recapitulationists, such as Serres and Meckel, from their later followers, most notably Haeckel, was an inability to suggest an effective mechanism for recapitulation. To some degree this may have been due to philosophical pressures of the day. Perhaps it was safer to go along with the easy option of falling back on the great Creator as the formulator of such phenomena, for as Agassiz (a noted "anti-Darwinian") wrote in his *Essay on Classification* in 1857:

> There exists throughout the animal kingdom the closest correspondence between the gradation of their types and the embryonic changes their respective representatives exhibit throughout. And yet what genetic relation can there exist between the Pentacrinus of the West Indies and the Comatulae, found in every sea; what between the embryos of Spatangoids and those of Echinoids . . . what between the Tadpole of a Toad and our Menobranchus; what between a young Dog and our Seals, unless it be the plan designed by an intelligent creator.

The publication in 1859 of Darwin's *Origin of Species* meant that for the first time there was an evolutionary framework against which the theory of recapitulation could be tested. Haeckel clearly grasped this in putting recapitulation in an adaptationist framework. So too did Agassiz. Even though he himself was not an evolutionist, he promoted the fossil record as a testimony of the ubiquity of recapitulation. But one person who really didn't stress the importance of recapitulation was Darwin himself. Indeed, Darwin seems to have been remarkably equivocal about the whole question of the importance of embryology to evolution. While, on the one hand, he wrote in the *Origin of Species* (6th edition) that "embryology will often reveal to us the structure, in some degree obscured, of the prototypes of each great class," he devoted a mere 10 out of 429 pages to the role of embryology in evolution. Perhaps Darwin was merely reacting against what had been the all-pervasive influence of embryological studies to species interrelationships in the decades leading up to the publication of the *Origin of Species*. Yet he clearly recognized the importance of embryonic development to classification, noting that "community in embryonic structure reveals community of descent."

On the surface it would seem that there is little doubt that Darwin accepted the recapitulationist point of view, for he states: "As the embryo often shows us more or less plainly the structure of the less modified and ancient progenitor of the group, we can see why ancient and extinct forms so often resemble in their adult state the embryos of exisiting species of the same class. Agassiz believes this to be a universal law of nature; and we may hope hereafter to see the law proved true."

Yet I think that maybe the reason why Darwin made so little of the relationship between embryonic development and his natural selection was that he may well have harbored serious misgivings about recapitulation as an all-embracing law. His unease is reflected in his observation that "the law will not strictly hold good in those cases in which an ancient form became adapted in its larval state to some special line of life, and transmitted the same larval state to a whole group of descendants; for such larvae will not resemble any still more ancient form in its adult state."

Perceptive as ever, Darwin was here presaging views that were to rise to dominance some seventy years later. Even so, recapitulation remained a potent and influential theory for the remainder of the nineteenth century, particularly among paleontologists. In terms of evolution, recapitulation was seen as functioning on the basis of two fundamental assumptions. First, "evolutionary change occurs by the successive addition of stages to the end of an unaltered, ancestral ontogeny"; second, that "the length of an ancestral ontogeny must be continuously shortened during the subsequent evolution of its lineage." These have been called "the principle of terminal addition" and "the principle of condensation," respectively. Much of the debate surrounding recapitulation in the latter half of the nineteenth century centered around three biologists, Ernst Haeckel, Alpheus Hyatt, and Edward Drinker Cope, each of whom published a major work on the subject in the same year, 1866.

Haeckel's influence was far-reaching, extending not only through biology and paleontology but also to aspects of politics, sociology, and religion. In their most extreme form, his statements came to be associated with concepts of "racial purity" and other doctrines of national socialism that emerged in Germany in the early part of this century. To explain the principle of terminal addition, Haeckel drew on Lamarckian views of the heritability of acquired characters that were more likely to be expressed in the adult stage. The principle of condensation was explained by Haeckel and his contemporaries as occurring either by an acceleration of developmental rate later in ontogeny or by the excision of certain stages from the developmental sequence. This enabled the remaining stages to be passed through more rapidly and made room for the terminal additions.

Haeckel coined the term *palingenesis* to describe the repetition of past evolutionary stages in developmental stages of descendants. He termed exceptions to palingenesis, that is, where different features were introduced into the developmental sequence, *caenogenesis*. Haeckel clearly delighted in making up new words and, as we shall see, this has been another reason for heterochrony's bad press over the years. Among the many new terms that Haeckel introduced (which included *ecology, ontogeny,* and *phylogeny*) was *heterochrony*. Haeckel used this last term to explain the displacement in time or the change in order of succession of particular organs, such as reproductive organs. Nowadays the term "heterochrony" is used in a more general sense to describe changes in the timing or rate of developmental events, relative to the same events in the ancestor.[4]

The American dinosaur hunter extraordinaire, Edward Drinker Cope, known these days more for his bitter rivalry with fellow dinosaur hunter Othniel Marsh than for his studies of evolution, became a fervent adherent of the cause of recapitulation. Perhaps because of his high public profile, Cope, along with fellow paleontologist Alpheus Hyatt, afforded to recapitulation a status that it had never previously enjoyed, nor was ever to achieve again. To Cope, a fearsomely prolific writer, who published a staggering fourteen hundred works during his life, species represented the modification of structures already in existence. New genera, on the other hand, arose, he believed, by additions to *or* subtractions from the ancestral ontogenies. Cope thought that recapitulation occurred by an acceleration in individual development, with ancestral growth stages being repeated at successively shorter and shorter intervals. A combination of terminal addition and acceleration would result in recapitulation.

Cope recognized, as had Darwin, that there were cases whereby there was less development in a descendant form. Here the later developmental stages would have failed to realize their potential and were deleted. Cope expressed these two facets of "more" and "less" growth as his laws of "acceleration" and "retardation," respectively. Where Cope and Haeckel were at odds over the question of recapitulation was over the extent to which the organism was affected. Haeckel saw recapitulation as primarily affecting the *whole* organism, whereas Cope with more prescience saw it as operating on *individual* organs. Even though he accepted the validity of both acceleration and retardation, Cope, like his contemporaries, saw acceleration, producing recapitulation, as the more important evolutionary force.

Cope's preoccupation with the species concept took a most unexpected turn in 1994. Unlike most mortals, whose corporeal remains are either consigned to the earth or incinerated, Cope had a much more imaginative idea for the deposition of his remains—he had his bones put in a cardboard box. Cope must have considered that he was the quintessential example of all that was human, for the method behind his apparent madness was that his body would be used as the type specimen of the species *Homo sapiens.* However, Cope, as University of Pennsylvania specimen number 4989, languished forgotten until recently. In 1994 the paleontologist Robert Bakker was instrumental in getting Cope's last wish fulfilled, for Bakker proposed Cope's remains as the type specimen of *Homo sapiens.* An eccentric paleontologist now stands as the embodiment of all that is human.

The Decline and Fall of Recapitulation

What caused the decline in the concept of recapitulation after its having been a major factor in evolution for so long? To a large degree it was one individual who, more than any other, contributed to its decline in the twentieth century—the American paleontologist Alpheus Hyatt. The extreme views that he held on the biogenetic law not only influenced many of his contemporaries but were also to play a crucial role in leading to the demise of the biogenetic law. On 21 February 1866, Hyatt read a short communication to the Boston Society of Natural History, which was to profoundly influence his own works and those of his contemporaries well into the twentieth century. As it is so short, I shall reproduce here in its entirety a contemporary account of Hyatt's talk.

> Mr. A. Hyatt made a communication upon the agreement between the different periods in the life of the individual shell, and the collective life

of the Tetrabranchiate Cephalopods. He showed that the aberrant genera beginning the life of the Nautiloids in the paleozoic age, and the aberrant genera terminating the existence of the Ammonoids in the Cretaceous Period, are morphologically similar to the youngest period and the period of decay of the individual; the intermediate normal forms agreeing in a similar manner with the adult period of the individual. He also pointed out the departure of the whorl among aberrant Ammonoids from its complete development among the normal forms, its final appearance as a straight tube in the Baculite, and the close connection between this morphological degradation of the whorl and the production of the degradational features in the declining period of the individual, demonstrating that both consisted in the return of embryonic or prototypical characteristics of the form, and partly of the structure.

Hyatt was firmly of the belief that evolutionary series of species passed through a "life cycle," much like that of an individual. He particularly stressed the importance of "senile" characters, these becoming the normal adults of tomorrow. Like Cope, Hyatt also explained recapitulation by acceleration. However, whereas Cope considered that *both* acceleration and retardation played a part in the theory of recapitulation, Hyatt's tunnel vision saw only acceleration. Hyatt's attitude, perhaps more than any other, was the undoing of Haeckel's biogenetic law. In attempting to explain all evolutionary changes as having being produced by acceleration, Hyatt went through some extraordinary mental gymnastics. To him, if what appeared to be a juvenile feature was retained into a descendant adult stage, it was explained as being only just a deceptive gerontic condition that had appeared beyond the normal adult stage (which had, of course, been conveniently lost) and which mimicked the ancestral juvenile stage.

The effect that Hyatt's ideas had on his contemporaries was really quite profound.[5] The classification of the Trilobita proposed by Charles Beecher, for instance, was based on the assumption that the ontogeny of later trilobites encapsulated the early history of the entire group. Thus, Beecher considered that "the process of acceleration or earlier inheritance has pushed forward certain characters until they appear in the protaspis, thus making it more and more complex."

Beecher viewed brachiopod evolution in a similar light, even though he recognized that there were many examples of "retardation" in brachiopods. Likewise, other groups, such as bivalves and echinoderms, were viewed in this same recapitulatory light. Yet, as science passed into another century, the first chinks began to appear in the armor of the extreme recapitulationists.

The cat was really put among the paleontological pigeons in 1901 by a Russian paleontologist, A. P. Pavlov. Unlike Hyatt, who had largely based his views of ammonite evolution on studies carried out on museum collections, Pavlov was a field geologist *par excellence*. By undertaking careful stratigraphic collecting of the ammonite *Kepplerites*, he showed conclusively that new characters often appear first in young stages and then spread to the adult stages of later forms. In other words, descendant adults show characters present in the *juveniles* of their ancestors, not the other way round. Not only did the early part of the twentieth century see the statue of Louis Agassiz turned on its head, tipped into this ungainly position by the effects of the great San Francisco earthquake in 1904, but it also saw the biogenetic law and Agassiz' beloved recapitulation undergo similar cranial inversions. As Pavlov wrote, "It is to be hoped that, under the influence of the facts, the limitations of the recapitulation hypotheses will soon be realized and that outside those limits the field will be left free for other interpretations."

Pavlov also documented similar phenomena in fossil belemnites, gastropods, and vertebrates. But Hyatt's stranglehold on ammonite workers was excruciatingly tight, to such an extent that even during the first two decades of the twentieth century English paleontologists, such as S. S. Buckman, A. E. Buckman, and L. F. Spath, formulated much of their ideas on ammonite evolution in terms of recapitulation.

From what I have told you, it would seem that I am implying that when it came to the biogenetic law most paleontologists during this period were short-sighted fools who saw nothing but the effect of recapitulation on every fossil they peered at. Well, this is probably, to a large degree, true. Yet there were serious dissenters, even among those who have always been unjustly lumped with the extreme recapitulationists. James Perrin Smith, professor of paleontology at Stanford University, was one. Although always tarred with the same brush as the recapitulationists, when one actually reads some of the original papers that Smith wrote on this subject at the time, it becomes clear that he was, in fact, an ardent and committed adversary of the recapitulationists.

In a paper that he wrote in 1914 called "Acceleration of Development in Fossil Cephalopoda," and which, since its publication, has been largely ignored, Smith pointed out that in ammonites *retardation* of development was in fact a more common phenomenon than recapitulation. His concept of the biogenetic law was, like that of Cope, much broader in scope than the rather restricted version espoused by Haeckel. It was certainly far wider than Hyatt's narrow-minded view. In what must rank as one of the most scathing attacks ever published in a paleontological paper, he castigated his fellow paleontologists for their "over-zealous" acceptance of Hyatt's ideas. As he wrote:

> It may be that, when this paper is read by ardent members of the "Hyatt school" of paleontologists and adherents of the biogenetic law, they will be inclined to call the writer a deserter from the camp, and to suggest that the paper ought to have been entitled, "Why recapitulation does not recapitulate." The writer is still a firm believer in the biogenetic law, but that law is not such a simple thing as it was once thought to be. In the youth of every theory everything is beautifully clear, and ideally simple. As time goes on we are compelled to drop one idea after another, until it almost seems that the whole will be lost. When sceptics concerning the recapitulation theory throw up to us that ontogeny does not always recapitulate, we are prepared to admit this, even to go further and admit that it does not often recapitulate. In fact, the writer would be prepared to go still further, and to state that, in the sense in which the term has been used by most adherents of the theory, it never recapitulates. Our over-zealous friends have claimed too much, and have done more to prevent general acceptance of the theory than a host of enemies.

Stern warnings, indeed. Yet despite Smith's urging, the domination of recapitulation, especially in ammonite studies, survived well into the 1930s, and its shock waves extended well into the 1970s. Surprisingly, despite the counterevidence from the fossil record, the sounding of the death knell for the biogenetic law came not from paleontological studies but from living organisms. Perhaps because paleontologists had carried the banner of recapitulation so high and for so long, it took much longer for them to release it from their grasp.

Thus, even though paleontologists like Otto Schinderwolf in Germany in the 1920s and 1930s were demonstrating countless examples of retardation (under the term *proterogenesis*), rather than recapitulation, it really took the dramatic statements of the English marine biologist Walter Garstang, working in quite a different area, to put the biogenetic law out of its, and everybody else's, misery.

After holding sway for more than one hundred years, recapitulation as an all-pervasive, all-dominant process was buried by Garstang once and for all. For what he proposed was quite the opposite. Rather than recapitulating phylogeny, Garstang argued that, in fact, ontogeny actually *created* it. For this Garstang proposed a new term, which will surface many times throughout this book—*paedomorphosis*. Literally meaning "child shape," paedomorphosis implies the retention of a childlike, or juvenile, morphology by descendant adults.[6]

The epitome of paedomorphosis —the axolotl or Mexican salamander, *Ambystoma mexicanum*

Other terms to describe such retention of juvenile characters by descendant adults had actually been coined in the nineteenth century, and as I shall discuss later, this profusion of terms is one of the numerous minefields one has to cross before being able fully to understand the role of heterochrony in evolution. For instance, in 1885 Julius Kollmann introduced the term *neoteny* to describe the retention of juvenile features in the axolotl *Ambystoma*, a salamander that reproduces as an aquatic larval form.[7] Usually found peering up at you from the bottom of an aquarium in your local pet shop, the axolotl, as well as having been a delicacy to the Aztecs, achieved fame not only from Garstang's promotion of it as a good example of paedomorphosis but because it was immortalized in verse by Garstang himself in what has become one of the most famous scientific poems. It was published in 1951 with a selection of his other poems called *Larval Forms with Other Zoological Verses*. Garstang's ploy in presenting complex ideas as simple poems was to try to help students understand more easily many of the fundamentals of biology that he was trying to teach. Paedomorphosis was eloquently expressed in verse in "The Axolotl and the Ammocoete":[8]

> Ambystoma's a giant newt who rears in swampy waters,
> As other newts are wont to do, a lot of fishy daughters:
> These Axolotls, having gills, pursue a life aquatic,
> But, when they should transform to newts, are naughty and erratic.
>
> They change upon compulsion, if the water grows too foul,
> For then they have to use their lungs, and go ashore to prowl:
> But when a lake's attractive, nicely aired, and full of food,
> They cling to youth perpetual, and rear a tadpole brood.
>
> And newts Perennibranchiate[9] have gone from bad to worse:
> They think aquatic life is bliss, terrestrial a curse.
> They do not even contemplate a change to suit the weather,
> But live as tadpoles, breed as tadpoles, tadpoles altogether!

Perhaps one of the reasons why Garstang's voice was heard, when others before his had largely been ignored, was because he saw paedomorphosis as the key to unlocking the mystery of the evolutionary routes taken by major groups of animals. Garstang's view, radical as it was at the time, was that backboned animals (vertebrates) may have evolved from something as seemingly inconsequential as the larva of a sea squirt (tunicate) by paedomorphosis. In other words, the early larval stage of some nondescript little sea squirt would, for some reason or other, have become precociously sexually mature. Peter Pan–like, it would have been trapped for the rest of its life in a state of perpetual youth. Sea squirt larvae were such attractive candidates as ancestral vertebrates because their free-swimming larvae were so different from the bloblike sessile adults. They had everything that an ambitious free-swimming vertebrate could ever have asked for: a notochord;[10] a dorsal hollow nerve cord; gill slits; and a propulsive tail.

Whether or not it was the novelty value of Garstang's proposal that so attracted the attention of other biologists, or whether it was its timely nature, coming, as it did, as the death knell was sounding for the biogenetic law, its impact was profound. For suddenly every man and his dog was being interpreted as having arisen from the retention of juvenile features found in some long-lost ancestor. I shall talk about paedomorphosis in man's dog in Chapter 4. In dog's companion, himself, the effect of paedomorphosis was championed by Louis Bolk, professor of human anatomy at Amsterdam, in the development of his ideas of *fetalization* in human evolution. Bolk believed that many characteristic adult human features were the product of retained juvenile primate features. He cited, for example, a flat face, lack of body hair, form of the external ear, loss of pigmentation, structure of the hand and foot, form of the pelvis, plus many more. Bolk was not a particularly strong supporter of the Darwinian views of evolution. Rather than extrinsic factors directing evolution he saw intrinsic forces, arising from developmental changes, as being most important. Writing in 1926, he stated: "Evolution is for organized nature what growth is for the individual: and (for the former) as for the latter, outer factors have only a secondary influence. They can never play a creative role, only one of modelling what is already there."

Although this paedomorphic view of man's origins has held sway in recent times, a small number of scientists, myself included, consider that this is quite an incorrect, and misleading, interpretation, as I discuss in the final chapter.

Recapitulation —R.I.P.

From the 1930s until the late 1970s, one of the most dramatic changes in emphasis ever seen in biology occurred. The pendulum swung from one extreme to the other, from an adherence to recapitulation to a promotion of paedomorphosis as the dominant factor in the relationship between ontogeny and evolution. It could be argued that paedomorphosis acted in many ways as a catharsis to the previously omnipresent concept of recapitulation. The Grand High Priest of this cleansing was Gavin de Beer, at one time (1950 to 1960) director of the British Museum (Natural History). In his two delightful books *Embryology and Evolution*, published in 1930, and *Embryos and Ancestors*, first published in 1940, de Beer set out to demonstrate the all-pervasive influence of paedomorphosis, particularly in the animal kingdom. He saw it as explaining the evolution of many groups of animals, including numerous types of invertebrates, including insects and some crustaceans; many vertebrates, such as man, chordates as a whole, flightless birds, and flying fish; and many fossil groups, including some ammonites, graptolites, and trilobites. As de Beer wrote in his chapter dealing with recapitulation in *Embryos and Ancestors:* "If Haeckel's theory of recapitulation had been correct, this chapter would be the longest and most important in this book. Instead, only a few scrappy instances can be found, and this mode has only played a minor part in evolution."

Recapitulation was dead.

While it might be thought that de Beer's books would have heralded a renaissance in the role of embryology in evolutionary theory, they failed dismally to have as much influence on modern evolutionary theory as might have been expected. Why this was so is rather complex. Clearly many biologists and paleontologists had completely lost faith in the whole concept of heterochrony following the demise of the biogenetic law. Another reason was probably the rise in the 1930s of genetics as a powerful tool in evolutionary theory. Here, it seemed, was the answer to the role of intrinsic factors in evolution, not developmental changes.

Another factor that relegated the study of heterochrony to the backwaters of biology was undoubtedly the coining of the most horrendous terminology ever perpetrated on mankind. I won't burden you with all of them, but when you realize that terms existed such as *paedomorphosis, peramorphosis, paedogenesis, palingenesis, proterogenesis, progenesis, pangenesis, phylembryogenesis,* and *prothetely,* plus many more, it's hardly surprising that most biologists and paleontologists steered well clear of the whole subject. I sometimes wonder if J. M. Barrie, the creator of Peter Pan, the child that never grew up (a classic case of paedomorphosis if I ever saw one), wasn't, either wittingly or unwittingly, poking fun at such terms with his choice of the main character's name—Peter Pan.

Gavin de Beer contributed to this confusion to some degree in his overtly complicated system of eight categories of heterochrony: caenogenesis, adult variation, neoteny (and paedogenesis), hypermorphosis, deviation, retardation, reduction, and acceleration. Now, if all this jargon seems overwhelming, I can certainly sympathize. But rest assured, most of these terms will not appear again (although a few which *do* help in explaining heterochrony will creep in every so often).

During the last two decades the study of heterochrony has undergone something of a renaissance. One book, more than any other, has been responsible for this upsurge in interest—*Ontogeny and Phylogeny*, by Stephen Gould of Harvard University. Published in 1977, this book has largely laid to rest the conflicting opinions over whether recapitulation or paedomorphosis has been the dominant process in evolution. For what Gould has argued is that the protagonists on both sides were right. There is no reason why recapitulation should be any more common than paedomorphosis or vice versa. Thus, Garstang's and de Beer's vision was as extremist as Haeckel's, and equally narrow.

What Gould also attempted was to try to sort out the tangled nomenclatural mess. Together with colleagues Pere Alberch, David Wake, and George Oster, Gould showed in a paper published in 1979 that between ancestor and descendant, development can either be reduced (resulting in paedomorphosis) or increased (resulting in what they termed *peramorphosis*).[11] This term essentially replaces the term "recapitulation," partly because of the particular extremist connotations associated with the term, but also because recapitulation, strictly interpreted, implies growth by terminal addition, whereas peramorphosis implies an increased growth, but not necessarily by terminal addition. From the nomenclatural morass Gould and his colleagues managed to rescue just six terms that they believed described all forms of heterochrony in which shape changes occur. Both paedomorphosis and its complementary phenomenon, peramorpho-

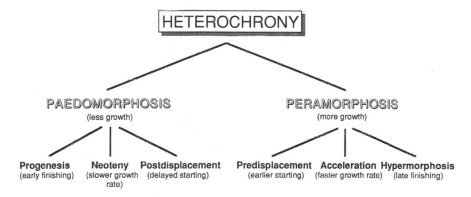

Hypermorphosis

A hypothetical beast. During its ontogeny it undergoes a number of obvious morphological changes. Peramorphic descendants develop "beyond" the ancestor; paedomorphic descendants retain juvenile ancestral features. In hypermorphosis, growth stops later; in acceleration, the horns and tail grow faster; in predisplacement, the horns and tail start growing earlier; in progenesis, growth stops earlier; in neoteny, the horns and tail grow at a slower rate; in postdisplacement, the horns and tail start their growth relatively later. Drawing by Sarah Long.

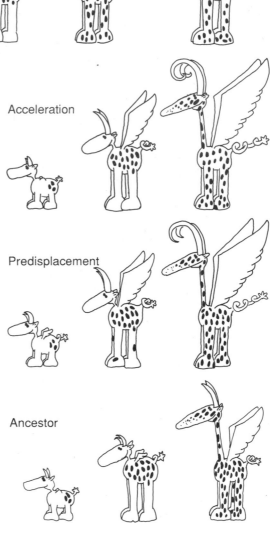

Acceleration

Predisplacement

Ancestor

PERAMORPHOSIS

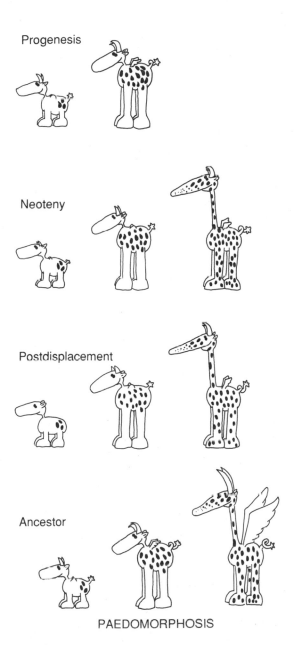

Progenesis

Neoteny

Postdisplacement

Ancestor

PAEDOMORPHOSIS

sis, could be produced simply by three processes. That is, organisms can develop either less or more than their ancestors merely by changes in the timing and rate of development.

Changes in *rate* of growth can produce either *acceleration*, when the rate increases, or *neoteny*, when there is a retardation in the growth rate. Acceleration (fast growing) will result in peramorphosis, that is, more shape change and an increase in complexity. Neoteny (slow growing) will produce paedomorphosis, that is, less shape change and a decrease in complexity. A change in the time that growth of an organ or structure starts development relative to others can occur either earlier (*predisplacement*) or later (*postdisplacement*). The early starters will result in peramorphosis, the late starters in paedomorphosis. The third way to change growth is by a change in the time of cessation of growth. Earlier cessation is known as *progenesis* (early finishing), producing paedomorphosis; later cessation as *hypermorphosis* (late finishing), producing peramorphosis.

These six processes therefore encompass all possible changes that can happen as an organism develops; they embody all that we think of as heterochrony. These six processes have effects on the shape and size of organisms, which we describe as paedomorphic or peramorphic. They can affect the whole organism, or different processes can affect different parts of the same organism. Some structures may even be affected by more than one process. The sum of such changes in organisms' morphology over time is evolution. So, whether we are looking at an animal, a plant, a bacterium, or a fungus, parts of it, or the whole of it, it can grow relatively faster or slower than its ancestor. Any part, such as a leaf or a finger, may start growth relatively earlier or later than in its ancestor. Likewise the leaf or the finger, or the whole organism, can finish growth earlier or later. And the same leaf or finger may grow faster or slower than in the ancestor. The magic of this is that ancestor and descendant can be like chalk and cheese, yet genetically they can be extremely similar. But what can be produced is the myriad of species that inhabit the planet today and the countless millions that have walked the Earth, swum in the seas, flown in the air, burrowed in the ground, or occupied any one of the millions of niches that have existed on the Earth for three and a half billion years.

A morphological consequence of these increases and decreases in the "amount" of growth that all these species show is that it can lead to the production of what are essentially either more complex, or less complex, structures in a descendant form, compared with its ancestor. A favorite animal of mine is the sea urchin. Common as fossils, especially forms like sand dollars and heart urchins that live buried in the sediment, urchins undergo great morphological and behavioral variation as they grow from tiny planktic, free-swimming larvae, into large spiky balls lurking in rock pools, whose main aim in life seems to be to stick their spines into the feet of unwary swimmers. But it is these self-same spines, and structures such as pores that pierce the hard, calcareous shells, that change through ontogeny and vary between species. So, in the case of a species that has evolved by the paedomorphic reduction in number of spines, we can consider that in this structure the descendant is simpler, and less complex. Likewise, if its rate of spine growth decreases and a smaller, less complex spine is produced, then this structure can be considered to show morphological simplicity. These forms are likely to have either fewer, and perhaps larger, cells, as I shall discuss in Chapter 7. In contrast, an urchin that has evolved more spines, or larger spines, is peramorphic and, consequently, morphologically more complex.

In a simplistic sense, therefore, we can consider paedomorphic species to be less complex than their ancestor, while peramorphic species will be more complex. Of course, defining complexity and simplicity in evolutionary terms is fraught with many dangers. For example, is a species that, overall, is more paedomorphic than its ancestor just more simplistic—and, consequently, does it show more simplistic behaviors? Probably not. Degrees of specialization and degrees of complexity are not necessarily directly correlatable. As I discuss in Chapter 7, some highly paedomorphic forms lead highly specialized, complex lives. Further, peramorphosis does not necessarily always equate with high levels of specialization. There are cases, particularly in trends toward increasing body size fueled by peramorphosis, and some of the more extreme versions of paedomorphosis, of high levels of specialization and consequently increased susceptibility to extinction.

Having broken the shackles of the past two centuries, and by adopting a simpler, more refined terminology, paleontologists and biologists are able to interpret the evolution of the living world in a much more unbiased fashion. Much of the research that has been undertaken in the 1980s and 1990s has involved testing Gould's hypothesis of the equitability of paedomorphosis and "recapitulation" (what is now called peramorphosis). Much of the rest of this book will be taken up with demonstrating the widespread occurrence of both of these phenomena among living animals and from the fossil record in order to show how evolution works from the inside. What this will show is that heterochrony is more than just a quaint evolutionary phenomenon deserving, as is usually the case, little more than a paragraph or two in textbooks on evolution. What I would argue is that it permeates every nook and cranny of evolution. Indeed, without it evolution wouldn't have happened. For it explains everything, from the shape of a delphinium flower, to a horse's foot, to the song of a bird.

Heterochrony operates at many levels. It occurs within species, resulting in the general variation that is present between individuals in populations—why, for instance, some individuals in a population of finches have slightly bigger bills than others; and why some leaves of a eucalyptus tree are slightly larger and of different shape from others on a neighboring tree. It operates between sexes and explains many

Even the Shell logo has undergone a sequence of more paedomorphic events in the minds of the ad men over the last century, evolving fewer ribs, so resembling more juvenile, less complex shapes.

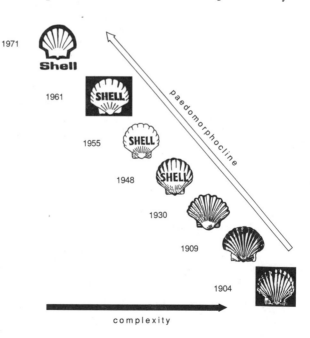

cases of sexual dimorphism, for instance, why some female spiders are much larger than males and why the males of some deer have much larger antlers than the females. As a central player in evolution, it plays a key role in the evolution of species and explains just how changes in the shapes and sizes of animals have occurred. It is also intimately related to the life history strategies of many organisms—in other words, how long they live, how long they exist as juveniles, how fast they breed, and how many offspring they produce. It is also becoming increasingly obvious that heterochrony plays a significant role in explaining behavioral differences between species. Any change, however subtle, in an organism's shape, size, or life style is likely to generate changes in behavior. Assessing changing frequencies of heterochrony through time can contribute to helping us understand the nature of great bursts of evolution that have punctuated the history of life. Its role in generating evolutionary trends is coming under increasing scrutiny, not only in explaining how and why certain groups of animals and plants have tracked down certain pathways, but also in showing how the agents of natural selection and heterochrony interact to generate directional evolution. For heterochrony is no replacement for natural selection—the two are partners in the evolution of life. Heterochrony is the logical link between, at one extreme, genetics, and at the other, natural selection.

3

The Shape of Things to Come

These little Limbs,

These eys and Hands which here I find,

This panting Heart wherewith my Life begins;

Where have ye been? Behind

What Curtain were ye from me hid so long!

Where was, in what Abyss, my new-made Tongue?

Thomas Traherne (1636–1674), "Dies Natatalis"

WHETHER IT OCCURS AS A CHANCE ENCOUNTER IN THE OCEAN, OR FOL-
lowing a few moments of unbridled passion, a sperm smashing its way
into an egg launches one of the most awe-inspiring events in the Universe—life. A helter-skelter proliferation of cell growth, division, and
differentiation leads to a living, respiring, feeding, reproducing structure that may comprise many billions of cells. A human body, for instance, undergoes an explosion of growth from a single cell at conception to something in the region of roughly 100 trillion cells. Yet of
this vast number only about 200 different types of cells are involved
in putting us together; and they are all found in other mammals, reptiles, and amphibians. The ordered growth of an animal, such as a human being, from a single cell to a politician, plumber, or paleontologist, replete with hair and nose, teeth and eyes, kidneys and liver,

typifies what has been going on on Earth, albeit to a lesser extent in many instances, for over three and a half billion years—an elegantly choreographed cascade of cell growth, division, and differentiation. But for all the remarkable scientific advances made over the last two centuries, we know precious little about the actual mechanisms of how organisms grow. After all, why should a mouse produce a tail, an eye, or a nose? How do the cells that make up the ear or the nose know where to grow? How do they know to produce an ear or a nose, when a cell from an ear can be the same as a cell from a nose? What is it that turns an initial, apparently featureless blob of an egg into something as remarkable as a living, feeding, squeaking mouse?

Building an Embryo

The growth and development of an animal like a mouse falls into three basic phases, called *neofertilization*, *differentiation*, and *growth*. An egg passes into the neofertilization phase just after it has been fertilized.

At this time most activity within the developing embryo is generated from information provided by the mother to the egg and passed by nurse cells. As cells undergo cleavage, the embryonic genes take over control of the operations, and both maternal and paternal genes are input. The time at which this occurs varies in different animals. In humans, this point occurs between the four- and eight-celled stage. This is a particularly critical period, for it is the time at which positional information is transmitted to the cell, telling it where it is in the embryo and how it should "behave" (in other words, whether it should differentiate, when it should divide, and when it should migrate). The distribution of this information is known as *pattern formation*.[1]

The initial maternal positional information dictating the position of cells in the early embryo and determining which end will develop as the head, and which the tail, can be derived from a number of sources. For instance, some mollusks use the point of entry into the egg by the sperm. In frogs, gravity can provide the reference point. In a number of animals, formation of the basic body plan takes place when the embryo is dividing into some 10,000 cells. In others, such as frogs, this takes place much earlier in development when there are far fewer cells. Positional information is distributed, and colonizing groups of cells that will develop into specific structures or cell types become established. It's something akin to a number of groups of workers being sent off to build a structure. Each group has its own set of tasks to accomplish in specific areas of the structure. While this establishment of founder groups of cells can occur in many groups of animals within just a couple of days of fertilization, in some "higher" vertebrates, such as humans, establishment of the basic body plan can take up to thirty days. Once the body plan has been established, each region of colonizing cells becomes progressively split into smaller modules. In other words, each of the initial groups of workers is split into smaller, more specialized groups of cells. However, wherever they are, the workers always carry with them the blueprint.

In recent years a great deal of study of such "modularization" has been carried out on the fruit fly *Drosophila*. It has been shown that an important role is played in modularization in *Drosophila* by a maternal gene calle *bicoid*. This gene plays a critical role in determining which end of the embryo will turn into the head, and which the tail. The *bicoid* protein demarcates the areas within the fruit fly egg. The head forms at the front end of the egg, where concentration of the *bicoid* protein is highest. A gradient of decreasing concentration of the protein from head to tail controls the overall pattern of segmentation. The variable concentration of the protein turns on genes in the early embryo which then generate more chemical gradients. These descendant gradients, following the activation of other classes of genes, eventually activate so-called *homeotic* or *Hox* genes. These are critical "switch" genes that control subsequent development of the fruit fly because they trigger the genetic pathway that establishes the identity of all the body segments in the embryo. The suites of homeotic genes that are switched on in each segment determine what the segments will look like and the structures that will develop on them, like wings or legs.

Amazingly, the gradients that switch on the homeotic genes in the fruit fly are actually active when the embryo is no more than a single cell. It was only in 1992 that it was demonstrated experimentally that the same system of chemical gradients is also present in vertebrates. Another important role that the *bicoid* protein plays in insects is the determination of the site of the junction between the head and the thorax. In insects there is a fixed number of segments in the head and the body. Likewise, the number of appendages that is produced: three

The genetic building blocks from which a fruit fly (and probably all other animals) are constructed. Each *Hox* gene constructs, in a highly ordered sequence, from front to back, various body segments. Changes in the regulation of *Hox* genes play an important role in evolution.

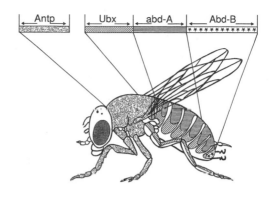

The same complexes of *Hox* genes that construct a fruit fly embryo also put together a mouse embryo. The numbers refer to different *Hox* genes.

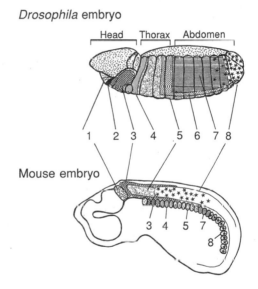

pairs on the first three segments of the body. While such basic structures form as part of a rigorously orchestrated program of development, experimentally it is quite easy to cause dramatic disruption to the program. If cytoplasm is taken from the front of the egg, rich in the *bicoid* protein, and injected into the middle part of the body, a head will grow out of the body. Such experiments highlight how profound structural changes can occur by tinkering with the developmental program. This also demonstrates the fundamental importance of the chemical gradients in determining which cells develop. Indeed, it is quite possible that such tinkering by natural selection in an ancestral arthropod many hundreds of millions of years ago may well have been the impetus for the evolution of insects in the first place (see Chap. 7).

One of the most important discoveries made in the last decade has been the unraveling of the genetic basis of the control of morphology. A wide range of organisms, from arthropods to vertebrates, share homeotic genes that determine body patterning. In arthropods these genes regulate segment morphology, appendage number and pattern, while in vertebrates they determine vertebral morphology, limb and nervous system pattern. Homeotic gene evolution, along with other developmental genes, resulted in the evolution of anatomical complexity during early vertebrate evolution. Yet the diversity within groups like vertebrates and arthropods has arisen largely by variations in the regulation of these genes, which are the ultimate arbiters of organisms' morphology. What links the homeotic genes of groups such as arthropods and vertebrates is their organization into gene complexes; their expression in distinct areas of the body in the same relative order along the anterior-posterior axis of the animal; and the possession of a sequence of 180 base pairs, known as the *homeobox* that encode a DNA-binding motif known as the *homeodomain*.[2]

The way this works in *Drosophila* is that there are eight homeotic, or *Hox*, genes in two complexes. Different *Hox* genes control the morphology of different parts of the body. One is known as the Antennapedia Complex, which targets the anterior part of the body; the other is the Bithorax Complex, which targets the posterior regions. The order of the genes on the animal's chromosome reflects the position of their expression in the developing animal. *Hox* genes do not build the animal, but control how it is built. Sean Carroll, of the University of Wisconsin at Madison, describes them as the "regulatory toolbox." They demarcate relative positions of structures in the animals, rather than control exactly what the structure will look like. The same *Hox* gene in two different species, however, may regulate the equivalent structure or region of the body in different ways. The last few years' research on *Hox* genes have revealed that they act as regulatory proteins and highlighted their promiscuous behavior in targeting a wide range of genes.

Carroll has argued that one of the ways *Hox* genes can influence morphological evolution is by changes in the position, timing, or level of their expression. Thus, many of the changes we see in terms of variations in numbers of body segments in trilobites or other arthropods, or in the number of vertebrae developed by different vertebrates, may be resolved with changes in the timing of activity of *Hox* genes. Here, perhaps, is the link between the morphological changes we see in the numbers of parts evolved in animals and their underlying genetic control. So in arthropods, homeotic genes control where particular appendages develop, as well as the type of appendage that occurs on a certain segment. As one example, the products of the Bithorax homeotic gene complex can repress a gene called *Distal-less*, thus inhibiting limb formation in a particular part of the body. The activity of such a suite of *Hox* genes some 400 or so million years ago may have been responsible for the reduction in number of limbs in a multilegged arthropod, resulting in the evolution of insects with just three pairs of limbs. The whole insect body plan, of head, limb-bearing thorax, and limb-free abdomen, is under the control of homeotic genes.

As well as suppressing limb development and being the prime movers and shakers behind paedomorphic reduction in parts, homeotic genes also appear to have the ability to modify original structures, so creating novel morphologies. Thus differences between, say, insects and crustaceans in the type and disposition of their limbs is likely to be firmly controlled by the different activity of suites of *Hox* genes in distinct parts of the body. It is interesting that the trends of segment and limb reduction that led to the evolution of insects parallel, on a gross scale, evolutionary trends in trilobite evolution. The earliest Cambrian trilobites possessed large numbers of body segments (up to 55 in the emuellids), and each segment is thought to have borne a pair of appendages. One of the general trends in trilobite evolution is for not only an overall reduction in the number of segments, and therefore limbs, but also a "developmental hardening" in segment number. Thus, early Cambrian species of trilobites possessed a relatively large number of segments; even within some species, this number varies. As we go up through the Cambrian Period, species and genera settle on a fixed number. By the time we get into the Ordovician and Silurian Periods, there are whole families, even orders, of trilobites that have not only fewer segments, but ones that are fixed in number.

We can argue not only for the activity of *Hox* genes very early in arthropod evolution, but also that their activity was pretty wayward early on, becoming more settled as the group evolved. From the evolving trilobites' point of view, such poorly controlled regulation early in their developmental history may have been a blessing in disguise, as it is likely to have led to the evolution of a wide range of genetically very similar, but morphologically quite distinct, morphotypes. Natural selection was given lots to play with. Interestingly, when I reviewed the role of heterochrony in Cambrian trilobite evolution, I found a preponderance of examples of paedomorphosis, reflecting, perhaps, the high activity of *Hox* genes, suppressing segment formation.[3] Post-Cambrian trilobites, on the other hand, which are more diverse morphologically, show more peramorphic features

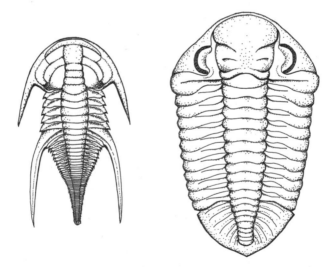

One of the earliest trilobites, *Emuella dailyi*, from 540-million-year-old Early Cambrian rocks in South Australia (*left*), had over 50 body segments, but the number varied within the species. *Acaste downingiae* (*right*), from 430-million-year-old Middle Silurian rocks in England, like all other members of the order Phacopida, had 11 body segments—no more, no less. "Developmental hardening" saw an improved regulation in development over time, and less wayward production of body segments.

affecting the sizes and shapes of existing characters. Such a shift in emphasis from one type of heterochronic process to another may therefore, in part, have its underlying cause in changing levels of homeotic gene activity.

Within a single group such as insects, the entire morphology has evolved under the action of the same set of homeotic genes: larval limb number, adult wing number, appendage morphology—all are regulated by homeotic genes. Although homeotic genes appear not to have played a part in wing development (as I discuss below), they were important in determining wing number. As Carroll has pointed out, these genes do not "make a wing" or "make a leg"; they merely modify existing developmental programs. So, in insect evolution, new sets of genes have not evolved; rather, new regulatory interactions have occurred between the homeotic proteins and the genes that actually participate in the structural formation of the limb or wing.

In vertebrates, similar rules appear to operate. Their segments are known as somites, and during development give rise to vertebrae. The actual numbers of vertebrae vary between different vertebrates. For instance, all mammals have 7 neck vertebrae (that's right, a giraffe has the same number as a mouse), whereas birds have between 13 and 25, while snakes have up to 454 precaudal vertebrae. There are different sorts of vertebrae, from head to tail: the cervical, thoracic, lumbar, sacral, and caudal. Transformation of one type to another is under the control of *Hox* genes, by loss of expression of particular *Hox* genes. As I discuss in Chapter 8, such evolutionary changes in *Hox* gene expression may have played a part in the evolution of turtles, for expression of particular vertebral morphologies is determined by the shifting domains of *Hox* genes up and and down the spine. *Hox* gene expression shifts with anatomy, not with somite level. Interestingly, if a particular *Hox* gene is set at the level of the forelimb at somite 17–18 in a chick, it will still be at the level of the forelimb in a mouse, but at somite 10–11. Carroll has suggested that one *Hox* gene group that demarcates the boundary between the thoracic and lumbar regions of the body may have been significant in the evolution of tetrapods from fish.

Patterns of development in living insects are, on the surface, very variable. For example, the biologist W. S. Haldane believed that God had an inordinate fondness for beetles, since there are so many varieties of them, but there is also a wide variety in their types of development. Yet underlying mechanisms are the same, differing only in the relative timing of the processes that promote segmentation. The differences in developmental patterns seen in beetles also occur in other orders of insects, in particular ants, bees, and wasps (Hymenoptera) and, to a lesser extent, grasshoppers and butterflies. The variability is expressed by the time in development that the segmentation is established, producing three broad categories known as *long-*, *intermediate-*, and *short-germ.*[4]

Long-germ embryos (e.g., the fruit fly *Drosophila*) have the body plan established by the end of the blastoderm stage (the early period of embryonic growth, when maternal genes dominate development). On the other hand, extreme short-germ embryos (e.g., the desert locust *Schistocerca*) form their body segments later in development after gastrulation (the period after the blastoderm stage, when genes of the developing embryo take over). Intermediate-germ embryos set down segments from the head to the end of the thorax by the end of the blastoderm stage; the posterior, abdominal segments are laid down after gastrulation. The long-germ pattern seems to be present in more evolutionarily "advanced" orders of insects. The evolution of segmentation earlier in development suggests that the timing of onset of the segmentation is displaced to earlier in development. Such "predisplacement" is one of the mechanisms that results in "more" developmental changes and a peramorphic, more complex, morphology. This occurs because the organism has, in theory, a longer time to undergo other developmental changes.

Embryologists have long understood that different species follow different blueprints, and that the timing of construction of the bodies will vary between species. Yet until the last few years our understanding of how such changes occur has been woefully inadequate. But the last decade has seen a revolution in our comprehension of some of the fundamental aspects of early embryonic development. Understanding mechanisms of growth allows the importance of heterochrony in evolution to be better understood. Changes seen between organisms that lived hundreds of millions of years ago can then be explained in terms of perturbations in the growth of cells that grew, functioned, died, then decomposed in deep time.

Exactly what is the nature of these chemicals whose concentration gradients are so critical to cell differentiation and patterning, and whose slight changes can cause such important evolutionary changes? Experimental studies carried out on a variety of animals, from fruit flies to chickens, and from frogs to mice, reveal that development is controlled, to a large degree, by the activity of molecules called *morphogens*. Originally suggested in conceptual form by the American geneticist Thomas Hunt Morgan (although not by that name), morphogens have been implicated as the controllers and orchestrators of growth, to a large degree from circumstantial evidence. Most telling is the fact that very early embryos if damaged are incredibly resilient. If tiny bits of an early embryo are displaced or even removed altogether, the embryo will compensate and develop normally. The same situation happens in twinning in mammals. Identical twins do not form, as popularly supposed, by the egg's splitting at the initial one- or two-cell stage. The splitting occurs at a later growth stage when the embryo has produced hundreds of cells. Yet two complete, undamaged individuals arise, even after such a profound schism.

Morgan's suggestion that morphogens exist (the term itself was introduced in the 1950s, by Alan Turing) arose from his work on earthworms—break one in two, and each half will regenerate all the missing bits to produce two individuals. What a mammal can only do when it is a few hundred cells large, an earthworm can do as an adult. Try splitting an adult mammal in half, and all you end up with is a very nasty, and very dead, mess on the floor. What Morgan proposed was that a chemical signal could give the embryo all the information its cells required for correct development. Embryologists today believe that information to direct how and where cells should grow is carried by morphogens that excite receptor molecules that are situated on the embryo's cells. These receptors then send signals to clusters of special genes that lurk in the nucleus of the cell. It is these genes that then direct the cells to where they should move, and when they should divide. Also, as a consequence through other interactions, the genes dictate when cells should turn into specific structures, such as bone or muscle. Clearly, for the orderly growth of a mouse the morphogens have to be produced at the right time, in the right place, and in the right amount, so that cell division, movement, or specialization can happen in a well-orchestrated sequence. Any perturbation in the timing of these events, or in the speed at which they happen, can have a profound effect on the developing embryo. The embryo will then follow a different developmental pathway that may lead to an adult that looks quite different from its ancestor.

One of the major breakthroughs in biology in the last decade has been the discovery of a likely candidate for the morphogen responsible for controlling segmentation. In both vertebrates and invertebrates, somatic growth occurs by the formation early in development of segments, formed sequentially, from front to back. Once established, these segments then develop their own peculiar morphological and functional characteristics. Thus, once the parts have been established at some rate, these parts will develop at a certain rate. Different parts may develop at different rates, so producing an organism with a variety of structures of different shapes and sizes.

The morphogen that plays a key role in segmentation is a substance called *retinoic acid*. Like many breakthroughs in science, the realization of the existence of retinoic acid and the role that it plays in morphogenesis was not the result of decades of careful, painstaking research, culminating in its discovery. According to Lewis Wolpert, in the Department of Anatomy and Developmental Biology at University College, London, who was involved in this discovery: "It would be nice to pretend that our discovery of the action of retinoic acid was based on many hours of deep thoughts. Not so. It came from the development of techniques for applying chemical substances to restricted regions of the limb bud, combined with inspired guesswork about the sorts of substances that might have effects."[5]

The unmasking of the remarkable ability of this substance to have such a profound effect on controlling growth was achieved by experimentally grafting a tiny bead soaked in retinoic acid into the front end of a developing embryonic chick limb bud. The result of this tiny incision and application of a minute amount of retinoic acid was the growth of an extra set of digits, in mirror image to the normal set. If less retinoic acid was applied, then fewer extra digits were produced. More recent work by Gregor Eichele and Christina Thaller at Harvard University seems to suggest that not only is retinoic acid present in developing limb buds, as the theory predicted, but that the concentration varies across the bud, with, as expected, a higher concentration in the posterior half.

As with all apparently delightfully simple explanations, further research is beginning to show that the system of morphogen-induced growth might not be quite as simple as it first seemed. One group of researchers, led by Nancy Wanek of the Rochester Institute of Technology, has suggested, for example, that rather than retinoic acid itself inducing growth, it might only be the agent that triggers the creation of a new polarizing region in which some other, unknown morphogen is operating. It is, of course, possible that this new polarizing region might occur by stimulating cells to produce even more retinoic acid. But such a scenario seems unlikely. Whether or not with the discovery of the effect of retinoic acid we have found the key that will unlock the secrets of limb development will only be confirmed by continued research. Only time will tell just how important retinoic acid is to the control of cellular development in growing embryos. The relevance of this to evolution is that as the factors controlling development are elucidated, so we will come closer to understanding the significance of developmental changes to evolution.

Deciding the Fate of the Cells

During the second phase of growth, groups of founder cells undergo division, differentiation, and migration to initiate the formation of specific structures in the burgeoning embryo. During this period, genetic information from the embryo, rather than from the maternal egg, becomes the dominant controlling force, directing the development of the growing embryo. This phase ends with the formation of particular organs and structures. In some groups of animals, positional information established in the neofertilization stage programs the ultimate fate of all the groups of cells. In other groups, such as echinoderms and vertebrates, cell differentiation occurs later. This means that cells within each founder group that are all ready and waiting to produce a particular structure are not of a specific type. They only take on specific tasks when they come into contact with other positional information which results in their differentiating. This means that with this type of development it is not necessary for each step to be carefully orchestrated; it just depends on where the cell is at any particular time. Shift its position in either space or time within the embryo and its developmental fate will be changed forever.

There is thus a great deal of plasticity in the regulation of cell dif-

ferentiation in the developing embryo in terms of how the cell re-
sponds to external stimuli. For instance, one cell, or a group of cells,
can act upon another cell to induce the formation of a third type of
cell. In order to do this, cells must have a mechanism for communication. The chemical environment in which the cells are developing is
therefore crucial to the ultimate fate of the cells. This has been clearly
shown by experimental work carried out by Jim Smith from the National Institute for Medical Research in London. He has shown that
just increasing the concentration of a specific molecule 1.5 times is
enough to induce cells to develop as muscle cells, rather than as skin
cells. The fate of the cell is therefore not under direct genetic control
from within the cell but is determined by factors operating in the environment *outside* the cell. One could almost place this within evolutionary terms by considering the cell's state as being bound up with
natural selection by external factors—where it is and when it is there
will determine what selective pressure it will experience and so what
it will "evolve" into. We come back once again to time as the great arbiter in controlling the cell's fate. If it arrives at its allotted site too
early, or if it misses the bus altogether, its fate will be changed, irrevo-
cably and absolutely.

One line of research is throwing up yet another contender for what
might be controlling cell growth—the investigation of the effect on
embryonic growth of the material that surrounds cells, the so-called
extracellular matrix. The extracellular matrix is the material within the
body that surrounds cells and that is made up of giant fibrous proteins
and globular glycoproteins. What is currently being argued, in particular by Betty Hass of Harvard University, is that interactions between cells and the extracellular matrix may have the effect of tripping switches nestling deep within the nucleus of the cell that will
activate genes. This idea has arisen from experiments such as one in
which epithelial cells from the mammary gland of a pregnant mouse
were grown in an extracellular matrix–free environment. The result
was that the cells grew flat and reverted to an ancestral condition.
Moreover, they didn't produce milk. However, when molecules from
the extracellular matrix were added to the mix, the cells reverted to
their original plump shape and organized themselves into sacs of cells
and began to secrete milk.

As well as providing a physical structure that allows the cells to adhere to the fibrous proteins in the extracellular matrix, when the cells attach it appears that they activate specific regulatory elements in the nucleus and so incite gene activity. The big question now is, how are these switches activated? David Ingber at the Harvard Medical School has shown that the cells must be stretched when they are attached to the extracellular matrix so that the switches can be triggered. He suggests that perhaps binding of the extracellular molecules to receptors on the cell triggers gene transcription by exerting an actual mechanical force on the cell wall and the nucleus. A case, perhaps, as Mina Bissell of the Lawrence Livermore laboratories in California has argued, of the extracellular matrix tickling the receptor, which tickles the cell wall, which tickles the nucleus.

During early embryonic growth, cells not only divide and move within the extracellular matrix, but, as I've explained, they also differentiate. The growth of a structure under such conditions can be considered to be dependent upon five basic factors:

1. The number of stem cells
2. The time of commencement of aggregation of cells
3. The proportion of cells that are actively dividing
4. The rate of cell division
5. The rate of cell death[6]

The activity of these five aspects of development will therefore control how the parts of an organism grow and what it will look like as a juvenile and as an adult. Any alterations in the timing of onset or offset or in the rate of any of these factors can have a great effect on changing what the animal or plant finally looks like, by changing the patterns of cell differentiation and growth. Alterations in developmental programs are therefore a potent force in evolutionary change. At the cellular level, this is achieved by simply changing either the timing or the duration of cell interactions, with resultant changes in size and shape of particular organs or structures, or in the establishment of new configurations. Such changes, particularly in terms of cell interactions, may even result in the establishment of new tissue types. Thus, changes in the timing or rate of cell activity may evoke major morphological novelties.

An Eye for

an Eye and

a Tooth for

a Tooth
A basic question often asked of evolutionary biologists is, how can a new feature evolve? We can argue that a structure, such as an arm or a tooth, can change in shape over countless generations, by changes in the rates and timing of growth, and that each change may confer some sort of advantage on the species. But how can completely new structures evolve? How, for instance, can something as wonderful as an eye ever have evolved? Can we really reduce it to the level of complex differential growth of cells? This certainly worried Charles Darwin. In a section in *The Origin of Species* entitled "Organs of Extreme Perfection and Complication," Darwin confessed that "to suppose that the eye . . . could have been formed by natural selection, seems . . . absurd in the highest degree."

As he argued, an eye is an eye is an eye. How could such an intricate structure evolve by small, gradual changes from an imperfect to a perfect state? Perhaps by understanding how cells migrate and grow early in development, we can propose a mechanism that is both plausible and realistic.

In a similar vein, if mammals evolved from reptiles, where did hair come from? How did the first backboned animals actually evolve bone? How did teeth evolve? One of the most exciting and fundamental questions that needs resolving in this whole debate is what effect variations in the timing and duration of interactions between different sets of cells very early in development contributed to evolutionary change. While we are still at a very early stage in our understanding of the mechanisms involved, research in these areas is beginning to unravel the mysteries of just how some of these major evolutionary events occurred. It is only by coming to grips with the essential rules of development that we will ever be able to understand fully the role of heterochrony in evolution: in other words, how changes in the timing of these cellular activities can affect the evolution of new species.

The eye, in many ways, provides a beautiful illustration of how the evolution of a specific set of epigenetic cascades has led to the formation of a very complex structure. In the early developing embryo, the forebrain grows into two regions: a central one, containing the cerebral hemispheres, and, on either side of this, two optic lobes. As the embryo grows further, these lobes extend toward the ectoderm and become the optic vesicles. As growth proceeds even further, they develop into optic cups. Within these cups, many different cell types differentiate: the neural retina, the pigmented retinal epithelium, and the iris.[7] When the optic vesicle has made contact with the ectoderm that overlies it, a portion of the ectoderm is induced to bud off as a structure that will become spherical and differentiate into a lens. This lens then interacts with the overlying ectoderm to induce a portion of the epidermis to become transparent and so become the cornea that overlies the lens—presto, an eye.

Selection acted on early vertebrates, at least 400 million years ago, to favor organisms in which this particular cascading sequence of events occurred. Being able to detect light is one thing; but inducing first a lens, then a protective, transparent covering, the cornea, would have been of tremendous selective advantage. The occurrence of fossilized "sclerotic" bones that encircled the eye of some of the most primitive fishes shows that this particular epigenetic cascade evolved very early in vertebrate evolution.[8] Had the last induction that produced the transparent cornea not occurred, then the eye would not have evolved as it did. Indeed, had any sequence in the cascade of eye development been displaced or not occurred at all, the eye would not have evolved.

Research that I am currently undertaking in a project led by my colleague John Long at the Western Australian Museum in Perth, in conjunction with Brian Hall, has revealed, rather surprisingly, that the early vertebrate eye, or at least the sclerotic bones that surround it, may have played a pivotal role in the evolution of jaws. The earliest known vertebrates were fishes known as the Agnatha. This means "without jaws." In his studies of early fossil fishes, Long realized that sclerotic bones that surrounded the eyeball are only found in one group of agnathans: the osteostrachans. And this group is generally accepted as being most closely related of all agnathans to the bony fishes (the gnathostomes).

It has long been realized that there is some sort of link between the bony ring that surrounds the eye and the tissue that forms lower jaws. For instance, in early chick embryos, mesenchyme can develop as bony rings around the eye when it comes into contact with the cells that will become the lower jaw. It seems possible, therefore, that the development of bony jaws may also be related to interaction between cells from the eye ring and the putative jaw. Central to this may have been the role of genes that inhibit cell condensations and scleral bone formation. Heterochronic changes in the timing of expression of such genes may have played a crucial role in determining the formation of bone in jaws.

Brian Hall of Dalhousie University in Canada has shown that changes in the interactions and positions of specific groups of cells within the developing embryo can explain how many structures form in animals and how such changes can provide a basic mechanism for generating evolutionary change. Much of the development that takes place in mammal embryos involves interactions between two major kinds of cells. One group is known as *mesenchyme,* the other as *epithelium.* The mesenchymal cells are a meshwork of isolated cells, and the epithelial cells are sheets of connected cells. These cell types are laid down at gastrulation, that stage when the effect of maternal genes is reduced and the genes of the developing embryo take over (the embryo at this stage is known as the *gastrula;* the maternal genes dominate during the *blastula* stage). Differentiation and growth begins at gastrulation. Here the embryo first starts to organize itself into its component parts. The nervous system begins to develop and initiates a sequence of inductions that will promote the formation of different cell types, tissues, and organs that will be present in the embryo, and ultimately in the adult.

Differentiation and growth of structures in almost all organs and tissues in developing vertebrate embryos occurs due to interactions between the mesenchymal and epithelial cells. Once an interaction occurs, it sets in train another, reciprocal interaction, which in turn sets another in motion, so establishing a cascade of interactions. These interactions are known as the *epigenetic cascade.* As the embryonic structures start to develop, interactions are initiated between adjoining systems—for example, interactions between nerves and muscles to develop muscles, and between muscles and skeletal elements for the formation of these latter two components. The first set of cascades results in the formation of the major regions of the body and triggers cell growth within specific organs. A second level of interactions occurs between these differentiated organs and tissues. Given that growth consists of these carefully orchestrated sequences of cascading cellular interactions, any slight changes, by a delay here or a speeding up there in an interaction, can have profound effects on the developing embryo.

Mesenchymal cells are generated from the mesoderm, one of three layers that make up the gastrula. The other layers are the ectoderm and the endoderm. The ectoderm will form external structures, such as skin and the central nervous system, which sinks into the developing embryo. Endoderm forms internal structures, such as the lungs, liver, and pancreas. Most of the rest of the structures, such as the skeleton, muscles, blood vessels, blood cells, heart, gonads, and kidneys, form from mesenchymal cells. These cells are very versatile and, depending on what they come into contact with, can produce a variety of tissues and structures. They may form nerve or pigment cells, or cartilage or bone, depending which epithelium they encounter. Any changes in the program that controls the timing of interaction of the mesenchymal cells and epithelium could greatly affect the basic components that are generated. An exception to this is the development of structures in the head. Lying initially in folds in the nervous system is the *neural crest*. Cells migrate away from the neural crest as the brain is growing, form mesenchyme, and contribute to other structures forming in the head.

In his research, Brian Hall has shown that changes in the timing of the interactions between major vertebrate groups, in addition to fishes, have greatly affected how new structures, such as jaws, have evolved in various groups. Most significant is the production of a cartilage known as *Meckel's cartilage*. This cartilage is important because it plays a principal role in initiating a whole cascade of events leading to the formation of bones, teeth, and, indeed, the whole mouth cavity. Timing of the interaction between the neural crest cells and the epithelium to form Meckel's cartilage varies between different major groups of animals. The interaction occurs later in amphibians than it does in birds, where the early interaction involves cranial epithelium. In amphibians the interaction occurs later as the neural cells migrate. It occurs even later in mammals, when mandibular epithelium is involved. This relatively late induction in mammals has been a key element in the evolution of bones in the middle ear, since it allows some mesenchyme cells to move their site of bone formation from the jaw to the middle ear region.

As I have said earlier, the evolution of new tissues or organs can occur by heterochronic changes in the timing of cell interactions. One of the most celebrated cases is the evolution of enameloid, in place of enamel, during the development of urodele amphibians, as they mature into adults. At metamorphosis, the epithelium of the tooth germ changes from participation in enameloid production to that of enamel. Enameloid is composed of proteins formed from both epithelium and mesenchyme. This is in contrast to enamel, which develops from epithelium, and dentine, which comes from mesenchyme. By a delay in differentiation of epithelial cells relative to mesenchyme cells, dentine is deposited after or at the same time as the epithelial proteins, producing enameloid rather than enamel.[9]

From these observations of the importance of the timing of cellular migration, we can see that the evolution of new structures is not a function of profound genetic mutations. Rather, the changing positions in space and time of groups of cells generate new structures and perhaps ultimately whole new groups of animals and plants. In an evolutionary context, we are dealing here with changes in the time that particular structures start to grow in the embryo, under the influence of changes in the timing of expression of certain proteins. In heterochronic terms, this is pre- and postdisplacement. Predisplace the onset of movement of a group of cells, and growth will either start earlier or generate a new structure. Likewise, postdisplacing the cells, both in space and time, can induce quite different morphologies. These are powerful evolutionary mechanisms.

Making a Wing

The evolution of the first wing probably occurred during the late Carboniferous Period, some 330 million years ago, when an insect took to the air. We know from fossils that forests of giant horsetails and ferns hummed with the drone of cruising dragonflies, whose wings spanned up to 70 cm. The very first insects that evolved during the early part of the Devonian Period, some 50 million years earlier, however, were wingless. It is highly unlikely that these dragonflies were the first flying insects; yet the fossil record gives us no clues as to the nature of the first flying insects. Most probably they possessed small wings that were perhaps originally adapted for some other use. How and why the first wing evolved has long been something of a mystery, and what transformed wings from one use into another, something as fundamental as a mechanism for occupying the hitherto unconquered ecosystem of the skies, has been the subject of much debate. Recent research has been tackling this problem from both ends: the adaptive changes that selected for enlarged structures that could become wings; and the developmental mechanisms that produced the diaphanous wing of the dragonfly, the huge, painted wing of the butterfly, and the whining wing of the mosquito.

While some scientists have suggested that insect wings originally evolved as solar collectors, the most favored explanation is that they originally functioned as gills in an aquatic ancestor. Until recently, how such structures came to be transformed into functional, beating wings, capable of supporting the insect's weight has been conjectural. However, a study of stoneflies from the Adirondacks by James Marden and Melissa Kramer at Pennsylvania State University has cast new light on this conundrum. Marden and Kramer observed that stoneflies possess wings that are essentially too weak to allow them to fly as such but which are used like sails on a windsurfer. They argue that because stoneflies are such a primitive group (with the remains of their ancestors having been found in Carboniferous rocks), they provide a useful analogue for formulating ideas on the evolution of insect flight. By experimentally manipulating the size of the stoneflies' wings they found that, not surprisingly, the larger the wings, the faster they could skim along the surface of the water in a breeze. This, they believe, is how wings may first have been used for locomotion. So rather than a somewhat surprised insect waking up one morning in a misty Carboniferous forest having sprouted a pair of wings overnight, complete with muscles for flight, Marden and Kramer see the stoneflies' method of locomotion as a possible intermediary scenario in which structures that were originally used for aquatic respiration were coopted into another use: first, perhaps, as oars used to row on the water's surface, then as sails, as they increased in size, probably by an acceleration in growth rate. If muscles were present to aid in rowing, then the increase in size of the wing may have been accompanied by an increase in muscle size, eventually enough for insects to leave the water altogether and take to the skies.[10]

Mayflies, relatives of the stoneflies, provide some support for this argument. Stoneflies, such as the *Taeniopteryx burksi* that Marden and

Kramer studied, have tiny hairs on both the legs and wings in both adults and juveniles. These hairs are of practical use when the insect is floating on the water, but of no use when flying, as they cause too much drag. Juveniles of mayflies, which are also aquatic, have similar hairs and skim across the water. However, the adult mayflies, which have fully functioning wings capable of producing flight, have lost these hairs. So by a little bit of evolutionary give-and-take, by a combination of a peramorphic increase in size of the wings and a paedomorphic loss of hairs, flight was achieved.

A great deal of experimental work from a developmental perspective has been carried out in the past few years in an attempt to try and understand the basic mechanisms behind the development of structures such as insect wings. This research has been largely conducted on the fruit fly *Drosophila* and has centered on trying to unravel the mystery behind how cells grow and are arranged in the developing wing. Such work is demonstrating that during embryological development there are multiple steps in the formation of any structures, such as a wing or a leg. Each step in this developmental cascade is crucial to the next step, and any change, whether it be a delay or a premature initiation, can put the whole system out of step and severely disrupt the construction of the part. While the vast majority of such "programming errors" are severely deleterious, and fatal, the occasional "mistake" may, if it occurs at the right time and in the right place, be adaptively significant and preferentially selected, thus acting as a trigger to the evolution of a new form. This is what heterochrony, to a large measure, is all about: delays or advances in the otherwise orderly sequence of events. We can see the influence of such changes in the fossil record or by interpreting evolutionary relationships of living organisms. Yet if we are to understand fully what lies behind these heterochronic changes, it is to the minutiae of embryological development that we must turn in order to appreciate how evolution is driven from within.

Wings are first thought to have evolved from a segment of the leg. The earliest winged insects bore wings on every thoracic and abdominal segment, not just on one or two thoracic segments as they do today. So any investigation into the number of wings in insects today involves unraveling the mechanisms that induced a paedomorphic reduction in wing number, for very early developmental stages of even these early fully winged ancestors would have been wingless. Sean Carroll and his colleagues from the University of Wisconsin at Madison consider that the timing of expression of homeotic genes has played a central role in determining wing number. Rather than promoting wing formation, homeotic genes repress it. Carroll and his colleagues consider that wings evolved in a genetic environment where homeotic genes played no role. At different stages of evolution, their role increased and is expressed by repression of the growth of wings on a number of segments. The time of expression of these genes can determine whether or not they have an effect on wing development. For instance, one gene inhibits wing development if it is expressed in the embryo, but not if its expression is delayed until later in ontogeny.[11]

Timing is of the essence.

It is becoming clear that there are a few crucial molecules that play important roles in making sure that construction takes place on time and in the right place. One such in insects, and which plays a crucial role in both wing and leg formation, is a protein known as *hedgehog*. The secretion of this protein in specific parts of the limb acts to induce neighboring cells to secrete other proteins, and so keep the construction schedule on track. In insects like *Drosophila*, every appendage in the adult is subdivided into an anterior and a posterior section. These regions are precisely defined at the cellular level and are established very early in the formation of the insect by founder cells that establish themselves in these regions. Within the posterior cells a gene, known as *engrailed*, is activated. This operates to inhibit any intermixing between cells of the anterior part and the posterior part. Recent research, in particular by Konrad Basler at the Universität of Zurich and Gary Struhl at the Columbia University College of Physicians and Surgeons, has shown that the boundary region itself, between the anterior and posterior compartments, is vitally important to the development of the limb. What they have discovered is that interactions between cells of the two regions can lead to the actual synthesis or direct transport of particular molecules that organize the behavior of the cells. And this behavior is a function of just how far the cells are from the boundary.[12]

Three genes in particular, known as *hedgehog*, *decapentaplegic*, and *wingless*, encode proteins that may be part of this signaling process.

While *hedgehog* is activated within the posterior part of the limb, the other two are expressed in cells in the anterior half that are adjacent to the *hedgehog* expressing cells. Basler and Struhl suggest that *hedgehog* is the critical factor in that by secreting the *hedgehog* morphogen, posterior cells organize limb patterning. Furthermore, *hedgehog* operates indirectly by actually inducing the formation of *decapentaplegic* and *wingless* proteins in cells in the anterior part of the wing that are adjacent to the boundary between the two halves of the limb. Such research is revealing the actual genes that express proteins able to specify the fates of surrounding cells. Any mutations that occur, whereby *hedgehog* signaling does not transpire, will have a deleterious effect on growth of the limb.

Time of expression of *hedgehog* is also significant. The time at which the activity of *hedgehog* is needed to specify segmental cell fate (6 to 9 hours) is different from the time when *hedgehog* signals to *wingless* expressing cells (3 to 6 hours). Not only time is critical, but also the concentration of the morphogen. Variations in the amount produced will have repercussions on which cell type will form at which position. Significantly, similar systems of cell control are being identified in vertebrates. There, the protein that fulfills a similar role to *hedgehog* is known as *sonic hedgehog*. Patrick O'Farrell of the University of California at San Francisco has described the function of *sonic hedgehog* as being like "a step in the cascade of sequential subdivision." *Sonic hedgehog* also operates in other parts of the developing embryo, most notably in the notochord, the floorplate of the neural tube and the zone of polarizing activity. In each of these regions a different cascade of morphogen activity and cellular growth is set in motion.

While a decade or so ago relatively little was known about how structures such as limb buds developed, we have come a long way since then in our understanding of this basic developmental process. Now that we have identified some of the genes and their protein derivatives that orchestrate the construction of a highly complex limb, and now that we appreciate the importance of timing of expression of these factors, we can start to get a handle on how some of the differences that we see between species might have originated. For instance, experiments have shown that inhibiting the expression of *Hox* genes sometimes has the effect of reducing digit size, perhaps by determining the growth rates of the cartilaginous precursors of the bone. In contrast, experimental overexpression of one *Hox* gene has been shown to result in the formation of an extra digit in a chick wing bud. So here we have evidence of direct genetic involvement with both the initiation of novel structures and alterations in the growth rates of existing ones.

Growth of the Parts

Having, at some length, examined the phase of differentiation, when the fundamental tissues and organs are set down, we now turn to the last, and third, phase of development—the *growth phase*. This is the period following cellular and tissue differentiation; it mainly just involves an increase in the number of cells. The cells will have established their positions, recognized and interacted with their neighbors, and be starting the process of growth. At this stage hormones, in particular growth hormones, come into play, determining how fast and for how long the cells will grow and replicate. This phase begins in the human embryo at about 2 months. Both differential cell growth and cell death will be occurring. Early in embryonic development, cell growth far outpaces cell death. The differential between these two will vary between various parts of the body. It will also change during development, cell death becoming increasingly common later in development. Effective cessation of growth at maturity is in many ways a reflection of the attainment of a relative equilibrium between cell growth and cell death.

Whereas morphogens and cell/cell interactions have such an important effect in early embryonic growth, in later development hormones play the role of controllers and directors of growth. Morphogens and hormones differ in that the former are short distance molecules, having very localized effects, whereas the latter are long distance molecules, extending their chemical tendrils throughout an entire organ or the whole body. In mammals, the most important substance controlling growth is the growth hormone *somatotrophin*. Produced by the pituitary gland, this hormone is stimulated into production by a neurotransmitter growth hormone–releasing factor. Inhibition of its production is controlled by a substance called *somatostatin*. Retinoic acid has recently also been found to play a part in this process by controlling growth hormone production in the pituitary cells. It does this by acting on growth hormone gene expression. Changes in the rate or timing of production of the retinoic acid, hormone, or hormone activator can produce heterochronic change in the tissues. In other words, the extent to which they are produced, or the time that they are produced, can greatly affect organ or tissue growth.

The impact of hormones on growth has been vividly demonstrated by experimental work carried out by Brian Shea of Northwestern University and colleagues. What they found was that in dwarf mice the pituitary gland produced almost no growth hormone, whereas in giant "transgenic" mice abnormally high amounts of growth hormone were produced. These mice grew larger by growing faster than normal mice, reaching a much larger body size than normal mice in the same time. In other words, growth was accelerated when more growth hormone was present.[13] Shea has used this mechanism to explain the evolution of body size differences between chimpanzees and gorillas. Although much larger, gorillas do not grow for a longer period than chimps; they grow at a much faster rate, perhaps under the influence of relatively greater production of growth hormones. In heterochronic parlance, they are relatively peramorphic, the accelerated growth having taken the gorillas "beyond" the chimps in terms of the "amount" of growth. As we shall see in the next chapter, the same probably holds true for breeds of dogs.

In other animals variability in growth arises from differences in the growth period, rather than the rate of growth. It has been shown, for example, that in musk shrews size differences arise from variations in the periods of growth. The larger strains grow for longer, the male's rapid juvenile growth phase slowing down after thirty-four days, compared with just fifteen days in the smaller strain.[14] These two different ways of attaining a relatively larger body size (delayed onset of maturity—hypermorphosis; or increased growth rate—acceleration) are under the control of fundamentally different genetic and developmental mechanisms. It was the lack of appreciation for so long that different mechanisms could engender the same morphological effect of either paedomorphosis or peramorphosis that led to some confusion over what heterochrony actually is, how it occurs, and what effect it has. Furthermore, it hampered the realization that development itself could evolve.

Changes in the timing of cessation of growth in mammals are produced, in part, by changes in the timing of sexual maturation. At the onset of maturity, when the juvenile transforms into the adult phase, growth slows down appreciably. However, there is a bit of the chicken-and-egg syndrome here, because it has been shown that in some mammals, including humans, body size itself can affect the time of the onset of maturity. But certainly, the larger body size of our species, compared with most other primates, can be attributed to our longer juvenile growth period, as I discuss in some detail in Chapter 12.

It is not only overall body size that is under hormonal control. Differential growth of organs arises from variations in tissues targeted by growth hormones. For instance, insulin-like growth factor I, a substance that promotes cell division and which itself is stimulated into production by the growth hormone somatotrophin acting on the liver and other cells, has little effect on brain growth, although it does affect overall body size. However, insulin-like growth factor II, which acts earlier in embryonic development, increases cell division in the brain and body, because brain tissue is more responsive earlier in development. Changes in the timing or rate of production of such a growth factor can therefore have profound evolutionary consequences.

Long bone growth in mammals may vary due to local variations in frequency of cell division in the bone. On the other hand, variation in cartilage growth is determined by the time at which growth ceases. This usually occurs when the cartilage has become so thick that it outgrows its blood supply. Changes in cartilage size and shape will therefore arise from factors such as changes in nutrient supply or alterations in the sizes of the blood vessels supplying the cartilage. The evolution of cartilage of different size or shape can consequently occur without any direct genetic input. Modifications of structures can also occur by differential cell death. In vertebrates such as reptiles, for instance, formation of digits occurs by interdigital cell death.[15] Reduction in the rate of cell death in turtles, ducks, sometimes even humans, can induce webbing between fingers. This can be selected for under particular conditions, such as aquatic environments.

We still have a very long way to go before we fully understand the almost infinite complexity of developmental systems at the cellular level. But herein must lie the key that will allow us to unravel the

mysteries of evolution. We can only assume that the patterns of cellular development occurring in living organisms are a reflection of
the changes that have been occurring in organisms for over a billion
years. Many of the apparent evolutionary changes that have been
documented from the fossil record (and some of which I document in
later chapters) follow a pattern that can be seen in all living things—
evolution of the organism's entire life history, by an interweaving
panoply of increases and decreases in development of parts, to produce countless millions of wholes.

4

It's a Dog's Life

There are some dogs which, when you meet
them, remind you that, despite thousands of years
of man-made evolution, every dog is still only two
meals away from being a wolf. These dogs advance
deliberately, purposefully, the wilderness made
flesh, their teeth yellow, their breath astink, while
in the distance their owners witter, "He's an old
soppy really, just poke him if he's a nuisance," and
in the green of their eyes the red campfires of the
Pleistocene gleam and flicker . . .

Terry Pratchett and Neil Gaiman, *Good Omens*

LET US ESCAPE NOW FROM THE SHELTERED, QUIET WORLD OF CELLS,
genes, and hormones and go out to where we can see the impact of
heterochrony all around us: out into our gardens, where it is reflected
in the shapes and sizes of the birds that fly around us; in the worker
and soldier ants that honeycomb our gardens; in the bees that polli-
nate our flowers; and in the flowers themselves. We could go to the
zoo and see what impact it has had on the evolution of countless
species of monkeys; of deer; of bears; of elephants—perhaps, of
everything. We could take a trip to the races and sit in the grand-
stand, where we could see its product in the thoroughbreds pounding
down the lush turf on hoofs that are a product of one particular type
of heterochrony, operating for more than forty million years. And as
we sit there, soaking up the sun, the pervasive effect of heterochrony

may creep, unannounced, even from the newspaper that lies folded on your lap.

Cartoon Evolution

There can't be too many newspapers around the world that, at one time or another, have not carried the Peanuts cartoon strip. In most of these, Charlie Brown is upstaged by one of the more laconic characters in the cartoon world, a dog called Snoopy. Now, Snoopy can hardly be described as anyone's idea of an ideal dog. Certainly not if you are after some atavistic hound, "just two meals away from being a wolf," that you expect to plunge its fangs into the backside of some unsuspecting intruder; and not if you are after a graceful, agile canine capable of chasing the most athletic of cats. No, Snoopy lacks these attributes. This is not through any fault of his own, but because of the tyranny of his evolutionary history. Consider carefully the morphological characteristics that typify Snoopy—just what does he look like? His body is relatively small; he has extremely short legs compared with many other breeds of domestic dog; his head is remarkably large, in comparison with his overall body size, yet his muzzle is poorly developed. So what Snoopy really looks like is an overgrown puppy. All of his morphological features cry out that they are juvenile canine characteristics—he is a paedomorph. You may well wonder how on Earth it is possible to talk about the evolutionary history of a cartoon dog. It's quite simple, really. Just look at Snoopy when he first appeared and you will see a very different cartoon dog.

I'm sure you will agree that this dog is quite different from the Snoopy that I've just described. Here we have a fairly typical, happy-go-lucky dog, much as you might see any day of the week doing indescribable things to your newly planted prize hydrangea. But just consider for a moment all the changes that this dog went through as it grew from a cute, cuddly little puppy into this bundle of bristling canine energy. The changes in shape that such an average dog goes through from birth to adulthood are profound. Apart from an increase in overall body size, many of the proportions of particular features change considerably. Most dramatically, the legs become relatively longer; and while the head becomes relatively smaller during development, compared with the body as a whole, the skull undergoes a great elongation, changing from a near-rounded shape in pup-

Snoopy today (*right*): a
paedomorphic dog,
little more than an over-
grown puppy. Snoopy
(*below*) not long after he
was born in October
1950: a "normal"-
looking dog.
Snoopy: © 1958 United
Feature Syndicate, Inc.

pies to a long-muzzled shape in adults. The eyes become relatively
smaller—a common mammalian trait. Like the muzzle, the jaw similarly undergoes a great elongation in many breeds of dog.

Now, just imagine what would happen to our bouncy cartoon dog
if it had stopped growing when it was a puppy, perhaps because it had
become sexually mature much earlier than its immediate ancestor.
The premature cessation of growth would cause the earlier-maturing
adult to retain many ancestral juvenile characters. So what would our
dog have looked like? Yes, you guessed it—Snoopy as we know him
today, a paedomorphic dog. For the more "normal-looking" dog is
actually a very early representation of Snoopy, as drawn forty-seven
years ago by his creator, Charles Schulz.

And don't think for one moment that our glimpse into the world
of evolutionary cartoons ends there. A classic example, once again of

Mickey Mouse in the early 1930s (*far left*), looking in some amazement at his more juvenile-like namesake fifty years later (*far right*). "The Evolution of Mickey Mouse" © Disney Enterprises, Inc.

paedomorphosis, was revealed to the public by Stephen Gould. His choice? None other than the Disney icon, Mickey Mouse. In the best tradition of scientific deduction, Gould analyzed a sequence of Mickeys, from the very first one that appeared in *Steamboat Willie* in 1928 to his fifty-year-old descendant. By a series of careful measurements of Mickey's cranial vault (height of his head), head size, and eye size, Gould found that as a proportion of body size all these three parameters had increased over the years. By plotting the same figures for Mickey's nephew, Morty (who presumably has more juvenile characters than Mickey), Gould found that in these features Mickey evolved toward the juvenile characteristics of Morty. In other words, he became progressively more paedomorphic over time.[1]

Chihuahuas to Wolfhounds

While in the cases of Snoopy and Mickey Mouse we are dealing with what might be a subconscious desire on the part of the cartoonists to select for "cuteness," a trait that is of selective advantage to cartoon characters, these are more than just flippant examples of heterochrony. For the morphological changes that took place in these fictional evolutionary histories mirror many of the evolutionary changes that have taken place in the real world, in dogs and mice and in many other organisms. Indeed, a recent investigation into which characters humans consider beautiful in each other revealed that they are often those associated with juvenile features: large eyes; height of the cheekbone; small nose; and small jaw.[2] We are drawn to these juvenile mammalian traits probably as part of our strong nurture bond as a species with our young. In contrast, we are often repelled by the opposite, peramorphic traits: large body size; large nose; and protruding lower jaw. Just look at all the Disney cartoon villains who bear these traits—the Queen in *Snow White*, the witch in *Sleeping Beauty*, and Jafar the sorceror in *Aladdin*.

Having looked at evolutionary patterns in cartoon dogs, let us look at evolution in real dogs—evolution controlled by humans in the form of artificial selection. For what is just a single species, the dog (*Canis familiaris*) exhibits a dazzling array of shapes and sizes. You only have to consider how the many and varied breeds of the domestic dog differ from each other. Many differ in terms of body size, from minute chihuahuas and Pekingese to gigantic St. Bernards. Others differ enormously in terms of leg length, from squat corgis to gazelle-legged greyhounds. Face shape differences are huge, from flat-faced pugs that look for all the world as though they have run into a wall at high speed to Hound of the Baskerville–type Irish wolfhounds. What we are dealing with here is the potent effect of artificial selection of differences in the "amount" of growth that the dogs have undergone during their ontogeny, from fertilization to adulthood.

To show how important heterochrony has been in artificial selection, Robert Wayne, while at the Johns Hopkins University, compared how much the shapes of the skulls of domestic dogs varied both between breeds and within single breeds as they developed.[3] In the case of domestic animals, artificial tinkering with growth trajectories allows comparison of heterochronies to be made between different breeds. It is also instructive to compare how much the shape of the skull varies during growth between different species of domesticated animals and see how that relates to the variability that exists between breeds.

The study of comparative shape changes during growth is known as *allometry*. Basically this is a descriptor of shape change of a particular structure or organ during growth, in relation to another feature, usually overall body size. Good examples of this occur in humans. We all know (many women painfully so) that, compared with adult heads, babies' heads are relatively much larger in proportion to the overall size of their bodies. As postembryonic human growth proceeds, the head, even though it is still actually growing in size, is doing so less than the body as a whole—the trunk, arms, and legs are trying to catch up after their slow starts. So, the head is becoming relatively smaller during growth. This is known as *negative allometry*. Other structures, such as arms and legs, on the other hand, are relatively very small (and functionally largely useless) in small babies. But as growth proceeds they grow rapidly and attain their specific functions of manipulation and locomotion, respectively. They become relatively larger compared with overall body size through development. This is an example of *positive allometry*. If there is no effective relative change in shape of the structure during development, then it is said to be *isometric*. The so-called allometric coefficient, in other words, the measure of the extent of the allometry, need not remain constant during growth. It may increase, or it may decrease, so that a particular structure may undergo, let us say, a relative increase in size at a certain stage in its development, but not throughout the entire growth period. Alterations in the timing of these allometric shifts may also bring with them significant behavioral changes.

What Robert Wayne found in his study of dogs was that the range of morphological types in different breeds is far greater than that of

The great difference in shape between three different breeds of dog, the "most" paedomorphic at the top, the "least" at the bottom. The shorter the face, the more vaulted the cranium. These longitudinal sections also show the relatively larger cranial cavity in the "most" paedomorphic breed.

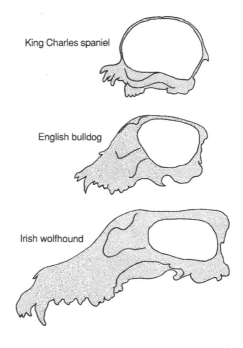

King Charles spaniel

English bulldog

Irish wolfhound

Proportionate differences in skull shapes of dogs and cats as they grow up. Dogs undergo much greater shape changes than do cats. This explains the greater diversity of skull shapes in different breeds of dogs, compared with cats.

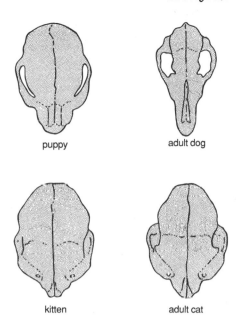

puppy   adult dog

kitten   adult cat

cats. In other words, even allowing for things like hair length and color, different dog breeds look more different from each other than do cat breeds. This difference has its basis in differences in the allometric growth of the skull. Wayne has demonstrated that skulls of domestic dogs grow with pronounced positive allometry in certain directions, while in cats skull growth in all directions is almost isometric. So, like Snoopy, the skull of a puppy is nearly spherical, whereas in many breeds, as the puppy matures, its head undergoes a great elongation to produce a long muzzle equipped with, more often than not, a fearsome array of teeth, all ready to be sunk into the softer part of the anatomy of something (or someone) the dog takes a dislike to. As a cat develops, on the other hand, the skull shows little appreciable change in shape.

During artificial selection, forms may be selected in which either the rate of shape change in the skull is increased or decreased or, alternatively, the existing rate of change is given either a longer or a shorter time to operate, by changes in the timing of onset of maturity. Then, not only is a breed of dog produced that may be very different in size, but it will also have a very different skull shape. Small dogs, such as chihuahuas or King Charles spaniels, retain a juvenile skull

shape as well as small body size, so they can be considered to be paedomorphs. This paedomorphosis will have been achieved either by a
neotenic slowing down of growth rate or by a progenetic earlier onset
of maturity, thus truncating the growth period. Larger dogs, like
Irish wolfhounds, Great Danes, and St. Bernards, on the other hand,
are examples of peramorphs that have developed "beyond" the average dog in body size and, to some degree, in head or muzzle size. This
can be achieved either by an acceleration of growth or by a hypermorphic extension of the growth period, brought on by a relative
delay in the onset of sexual maturity.

In contrast to domestic dogs, extending or contracting the growth
period of domestic cats or changing their rate of growth will have
little effect on the shape of the skull. This is due to the isometric nature of the growth of the cat skull. So, differences between breeds of
cats are more subtle. The most extreme differences in the skull, arising from small heterochronic shifts in timing of development, are
shown by Siamese and Persian cats. Thus, some long-faced breeds of
Siamese can be contrasted with some of the flatter-faced Persians,
probably due to a relatively more peramorphic growth of the anterior
region of the skull. Pigs also show strong changes in skull proportions
during growth, and like dogs, hundreds of breeds are recognized.
The effect of artificially selecting for different maturation times is
shown in breeds of rabbits. The small, paedomorphic "dwarf" rabbits
attain sexual maturity at about five months, whereas the huge Angora
rabbits do so at about nine months.

So, for all those dog lovers who are attached to their respective
breeds of pooches, let them give thanks to heterochrony. For without
it they would all be stuck with dogs of the same unprepossessing nature and appearance. (And what, you might well ask, would that do
for dog lovers' morphological variation, often seen as mirroring that
of their pet? It too would be severely reduced!)

Variety Is the Source of Life

Differences in appearance between individuals of animal and plant species that are the product of subtle
shifts in growth rates or timing of maturity may, like
those of the domestic dog, be very obvious. Set a King
Charles spaniel next to a Great Dane, and it becomes
pretty clear that artificial selection has targeted those

characteristics most affected by changes in the growth and development of these dogs. But what of natural selection, as opposed to artificial selection, acting on the normal variation that occurs within populations of all species? If we, as a species, are able to select such profound differences with great facility over a relatively short period of time, then surely "Nature," operating over millions of years, must have been able to produce similar changes. How does heterochrony interweave its magic with natural selection to produce the variability that we see today in many species, and which the fossil record shows us has been occurring for hundreds of millions of years?

One species that is the most morphologically variable is our own. In addition to such obvious traits as eye color and hair color, much of the morphological variation that occurs within humans involves differences in the shape and size of the body as a whole, plus variations in certain structures, such as limbs. As with that other mammal, the dog, these arise from variations in the rate of growth of particular body parts, or in the period that they grow. It is well known that the extent of growth is influenced, to some degree, by environmental factors, in particular diet. But such differences in development between individual humans that arise from differences in growth, or from inherent genetic differences affecting development, can all be considered as heterochrony. The rich variety of human types within single populations arises, in part, from the same factors that operate in dogs, namely the pronounced allometric growth of many of our skeletal features: arms, legs, jaws, feet, hands, and so on. Compounding this is one of the characteristic features of human development, one on which I will elaborate in Chapter 12—our very long period of pre-adult development. What this means is that any slight difference in growth rate between two individuals will be amplified over the long period of development.

Take a leg—any leg. Irrespective of whether the leg belongs to a male or a female, its constituent bones grow from a primordial limb bud between twenty-five and thirty days after conception, until shortly after puberty. Say we are looking at a fifteen-year growth period; then any minute change in growth rate between two individuals, or in the time that the limb starts developing, will affect the comparative lengths. Likewise, if the limb stops growing at different times, then the length could vary. One factor contributing to greater male height in humans, compared with females, is the relatively delayed onset of maturity, giving a longer growth period. If the relative difference in growth rate between a femur bone in two individual humans was the same as between two mice, the final difference as adults would be much more pronounced in humans than in mice, because of the longer growth period. It's like two people starting off from the same point at the same time. If their paths are slightly divergent at the beginning, then the longer they walk, the farther away from each other they get. Likewise, if they follow exactly the same path, but if one either starts earlier or stops later than the other, the distances that each one travels will vary. The same with the growth of any structure or organ. Perturb its rate or timing, and the end result will differ. More drastic changes in growth rate can effect quite spectacular differences in shape and size. Brian Shea of Northwestern University has shown that the paedomorphic nature of various groups of African pygmies, combined with their small body size, is a result of reduced levels of the growth hormone insulin-like growth factor 1 (IGF-1), particularly during the adolescent growth spurt. The same effect can be seen in poodles, lower levels of IGF-1 producing miniature poodles, and even lower levels of IGF-1 resulting in the even smaller toy poodles.[4]

What this all means is that where the growth of traits is allometric, any factors that influence their growth in comparison with the overall growth of the body are likely to have an obvious morphological impact. This in turn is liable to influence the functioning, and therefore the behavior, of the organism. The inevitable consequence of this is the direct influence of the organism's "fitness," in terms of natural selection. Common external environmental factors influencing growth in mammals, for instance, are litter size, age and condition of the mother, differential mortality of smaller young, and nutrient status. Basically, differences in size of any particular adult structure, which may affect shape, can be accounted for in the standard three parameters of size at birth, growth rate, and length of growth period, or any combination of these.

In assessing environmental factors influencing differences within species in size and shape of traits in the North American deer mouse, *Peromyscus maniculatus*, Ken Creighton and Richard Strauss of the University of Michigan studied the variability in brain size relative to body size resulting from heterochronic changes in growth patterns. They compared two named subspecies of *Peromyscus maniculatus* and found that they both follow the same growth trajectories for body weight and brain weight from birth to eight days. After that, the allometric relationships between the two subspecies vary from the tenth day through to the fortieth, due to differences in the time that cranial growth stops. The difference between the two subspecies in the time the brain stops growing is only about four days. As a consequence, one subspecies, called *Peromyscus maniculatus bairdii*, is paedomorphic with respect to the other, *Peromyscus maniculatus gracilis*, due to its brain having stopped growing earlier. It therefore has a shorter time to develop and ends up with a smaller brain that is slightly less developed and so less complex.[5]

Although it might be easy to casually dismiss the gross morphological differences that are induced by heterochrony as being little more than the effects of allometric scaling lacking any substantive genetic basis, the potential adaptive significance of such changes can be profound. For the evolution of a species requires genetic isolation, and the appearance of populations within a species that may have a particular morphological feature generated by heterochrony provides the necessary mechanism to achieve the genetic isolation. If the different structures result in the population behaving differently from the "parent stock," the population might achieve reproductive, and thus genetic, isolation. Such isolating mechanisms may also form if there is a change in the timing of onset of maturation, even though the resultant morphological change is minimal. This can be graphically demonstrated from what is essentially an example of incipient speciation going on at the moment in the North American apple maggot fly.[6]

During just the last 200 years the apple maggot fly, *Rhagoletis pomonella*, has evolved a number of races that have become adapted to feeding from a range of different trees. Originally adapted to feeding solely from hawthorn trees in North America, this fly has spread to infest a range of introduced trees, such as apples, cherries, roses, and pears. As early as 1864, a minister and eventual state entomologist for Illinois, Benjamin Walsh, observed that behaviorally, offspring of a fly from, say, an apple tree are more likely to lay their eggs on other apple trees. While this, in itself, is sufficient to engender isolation, there are also significant differences in the timing of onset of maturity of the fly on different hosts. This will further enhance the likelihood of genetic isolation. Under laboratory conditions, the "ancestral" hawthorn fly takes between 68 and 75 days to mature. The apple fly, on the other hand, takes only 45 to 49 days. Flies that infest fruit of the dogwood, however, take much longer, between 85 and 93 days. The reason that the onset of maturity is so variable is that it is timed to coincide with the period of maturation of the fruit of the host tree. The result has been not only the establishment of behavioral barriers to gene flow but also developmental barriers, since mating time of the flies from different hosts will vary. Given more time and continued isolation, it would be expected that genetic differences between the populations would increase to such an extent that even were they to be placed in a position to mate in the future, viable offspring would not be produced. So, what we have here is a number of species in the making—evolution taking place before our eyes.

The morphological changes that I have described as arising within species by the operation of heterochronic processess have all involved changes in the shape of structures that appeared, in some form or other, very early in development. However, in most groups of organisms there are a number of characters that appear at later stages of development, often in large numbers. For instance, in sea urchins, the characteristic spines are produced throughout the growing phase of the animal. It is possible, as I shall show in later chapters, to analyze evolutionary changes in the numbers of spines, because the rate of production of these structures can change; or, alternatively, the period during which they can be produced might vary between ancestor and descendant. We are dealing here with heterochrony of parts (so-called meristic characters), rather than heterochrony of shapes. In many ways, describing such heterochronic changes of parts is easier than describing shape changes, because all we have to deal with are numbers of structures produced in a particular time, rather than trying to quantify changes in a complex structure.

An example with which most of us are familiar (often painfully so) is the timing of eruption of our teeth, in particular our wisdom teeth. My own entire quota of wisdom teeth had erupted by the time I was about twenty years old. However, my wife's wisdom teeth have failed to erupt at all. For this character my wife can be considered to be paedomorphic compared with me. In an evolutionary context, the timing of tooth eruption in later hominids, such as *Homo sapiens*, has been progressively delayed. While I don't like to admit that my wife is evolutionarily more advanced than I am, one could say that in this character she is slightly further along this particular evolutionary road. (She has also even retained into adulthood one of her milk teeth—an extremely paedomorphic trait!) I shall discuss tooth development in hominids in greater detail in the final chapter (hopefully, in part, to extricate myself from the position of being considered an evolutionary laggard).

While all of the examples I have described (with the exception of the apple maggot flies) demonstrate the importance of heterochrony in generating morphological variation that might lead to behavioral differences, unless such variants are preferentially selected, there will be no evolution. So while I might seem to be belaboring the point through this book that these differences can arise from little genetic change, I am certainly not launching a challenge to the classical "Darwinian" mode of evolution. On the contrary: natural selection, in the classical Darwinian sense, comes into play as the partner of heterochrony, in being the factor that selects for the variants that arise from these slight perturbations in developmental programs. However, I must once again stress that heterochronic variability, such as that seen in domestic dogs, is, by virtue of the pronounced allometric growth, the very factor responsible for the success and adaptability of the species. Such variability arising from pronounced allometry of key structures can therefore be an important precursor of evolutionary change.

Insects— Getting a Head Start in the Sex Race

Variations within species that arise as a consequence of slight differences in growth rates or timing are just one way that heterochrony operates within species. As well as this "passive" type of selection (a sort of background evolutionary white noise), the environment itself can be much more proactive. Rather than merely passively selecting what has already been formed, environmental factors can actively induce heterochronic change. If we return to our restaurant analogy, instead of simply sitting back and eating whatever has been served, the customers will actually have a say in what they will eat and how the food will be cooked.

The extent to which the environment induces particular developmental strategies by triggering heterochronic change can be profound. Only certain members of a population may be targeted to produce unusual growth forms; or, alternatively, only specific parts of the body might be focused upon. One of the more striking examples is the influence that diet has on growth in the caterpillar *Pseudoaletia unipuncta* and, indeed, on the overall evolution of many different types of caterpillars and grasshoppers.

Experiments carried out by E. Bernays, working at the University of California at Berkeley, have shown that by feeding different caterpillars of a single species different diets, the proportionate size of the head can change quite dramatically, depending on what the caterpillars were being fed. What Bernays found was that those caterpillars fed a diet of rather tough grass developed heads twice the mass of those that ate a soft artificial diet. Not surprisingly, those fed an intermediate diet developed an intermediate head size. These differences in head size have been attributed to changes necessary to increase muscular development in the tough-grass eaters, which affect the overall head size. When Bernays analyzed eighty-two species of grasshoppers and seventy-six species of caterpillars from North America and Australia, she found that grass eaters consistently have relatively larger heads than herbaceous plant eaters.[7]

Rather than having to invoke a "Lamarckian" scenario (like giraffes getting longer necks because they were stretching higher and higher to the tops of the trees to get juicier leaves), there is a genetic basis for these changes in the heads of the caterpillars and grasshoppers. Again we see the partnership of heterochrony and natural selection working together: developmental changes producing some individuals that had larger heads, in the case of grass feeders, and so were preferentially selected. If you have tougher mandibles and can cope better with the particular diet, then you are more likely to grow bigger and be reproductively more successful. Here we have a nice example of heterochrony (in this case, peramorphosis, producing "more" growth) acting not on the whole organism but on one particular structure.

While such examples show how rate changes can induce hetero-chronic change in specific morphological structures, environmental factors that influence the time at which an organism attains maturity will affect the entire organism, both in terms of its size and shape and, probably, its behavior. In social insects there is one factor that can greatly influence the timing of onset of sexual maturity, and thereby affect the size and shape that the adult animal achieves. That factor is the production by some members of the communities of particular hormones called *pheromones*. These can trigger a cascade of physio-logical changes that either delay the onset of maturity (hypermorpho-sis), such as occurs in honeybees, or bring it on earlier (progenesis), as occurs in the desert locust *Schistocerca gregaria*.

It has long been known that the mere presence of adult males within colonies of these locusts when population density is high will result in the precocious onset of sexual maturity in any immature males or females with which they are living. Mature females also have a similar ability, but they are less effective than the males in activating maturation. But paradoxically, females that are less than eight days old secrete pheromones that have the opposite effect, having the capacity to retard male maturation. The selective advantage of this strategy would seem to be that the females do not want the males to be ready to mate before they are. The high population densities of-ten found with these locusts means that there is usually synchro-nous maturation, any laggards being accelerated by the merest whiff of a pheromone. The other advantage of this strategy is that it stops potentially precocious maturers from becoming mature at too small a size. It has been discovered that the pheromone that acceler-ates maturation is an extremely volatile substance that covers their bodies. A mere 1/5000th of the pheromone that is present on the body of any one individual is enough to excite a response in an imma-ture male.

Locusts are not the only insects to fall under the alluring spell of sex pheromones. Similar maturation-accelerating pheromones appear also to be present in ladybirds, taking effect when immature adults congregate. Synchronous onset of sexual maturity is followed by mat-ing, then by dispersal. Similarly, pheromones have been shown to regulate maturation in various species of ants, bees, and wasps. In these insects, the mere presence of the queen has the effect of inhibiting sexual maturity in the workers. If the queen is removed, then ovaries develop in the female workers and eggs are laid.

The presence or absence of other hormones at particular times can have a profound morphological effect in other insects. One of the classical life history cycles long taught in biology classes at school is that of the aphid. Aphids sometimes occur with wings, and sometimes not. Although generally not expressed in these terms, those forms without wings can be considered to be paedomorphic, because they retain the juvenile state of winglessness. These wingless forms arise because they have an earlier onset of sexual maturity than their flying counterparts. They are more fecund and have a shorter generation time than the winged aphids. Wing development is under the control of juvenile hormone (see below). It has been found that wingless forms have a larger and more active *corpora allata*. This is the organ that secretes juvenile hormone and suppresses the onset of maturity. This same juvenile hormone plays a crucial role in caste differentiation in termite colonies. A larval form, known as a pseudergate, can molt into any of three castes: winged imagos, replacement reproductives, or soldiers. Alternatively, they can molt without growing or differentiating. The formation of the different castes appears to be under pheromone control, again mediated through the juvenile hormone system.[8]

Soldier Ants and Assassin Bugs

Hormones can have a tremendous effect on growth in arthropods, such as insects and crustaceans. For here we can see how experimentally induced heterochronic changes can have a profound morphological impact and, by mimicking situations found in nature, provide an insight into just how changes between species, or even between different polymorphs (such as worker ants and soldier ants), can transpire by relatively simple changes in the timing of hormonal production. Environmental effects can act on the hormones and change the course of development of the organism.

Changes in temperature have been shown to act on animals' hormonal systems, in particular by affecting both the induction and inhibition of maturation. They do this by acting on the hormones involved with growth and development. Of particular importance is the hormone that controls molting. Known as *ecdysone*, this hormone induces molting in a number of groups of arthropods, including insects, crustaceans, and spiders. Neurosecretory cells in the brain produce a hormone that stimulates the synthesis and release of the molting hormone ecdysone, which then induces the molting process.

The other major hormone that controls growth is *juvenile hormone*. This has also been shown to act in a similar manner in both insects and crustaceans. Secretion of juvenile hormone by certain glands modifies the expression of the molt and acts in conjunction with the molting hormone. It has been suggested that juvenile hormone may act directly on genes that regulate development, so repressing their activity. While juvenile hormone is present at a sufficiently high level, the insect or crustacean will display juvenile morphological characteristics. When this hormone ceases to be produced, metamorphosis to the adult phase occurs. From an evolutionary perspective, any change in the timing of cessation of production of the juvenile hormone is likely to have a profound effect on the ultimate appearance of the adult, compared with its ancestor, for as an adult it will retain characters present only in ancestral juveniles.[9]

Temperature variations can also alter the molting sequence in arthropods and so induce heterochronic changes. Secretion of the hormone ecdysone activates the epidermal cells to secrete a new cuticle. So, any changes in the timing of induction of deposition of the new cuticle will affect the degree of developmental change the arthropod goes through between molts. Thus, if the cuticle deposition occurs soon after the previous molt, there will be little morphological change between molts. If, however, the onset of deposition of new cuticle is inhibited by delay in ecdysone secretion, there will be a greater degree of morphological development between molts. The classic work undertaken by Sir Vincent Wigglesworth in the 1930s showed that artificially inducing the bedbug, *Cimex*, to precociously deposit its cuticle results in the appearance of certain morphological features in an intermediate stage of development. Much the same results have been obtained by accelerating the onset of molting in the assassin bug, *Rhodnius*.

Wigglesworth's pioneering work on the effects of experimentally inducing precocious maturation in *Rhodnius* by artificially removing the source of the juvenile hormone revealed the great importance of this hormone to the timing of maturation and consequently what the adult would look like. Similar experiments on the silkworm, *Bombyx*, and the phasmid *Dixippus* have likewise shown how paedomorphs could be produced by artificially reducing the number of juvenile molts. A similar effect is seen in fossil trilobites that lived up to 530 million years ago: precocious onset of maturity—most likely caused by premature inactivation of juvenile hormone—resulting in fewer body segments being produced and ancestral juvenile features being retained in the descendants. So, by undertaking experiments that tamper with the developmental program of living animals, we are able to hazard reasonably informed guesses as to what evolutionary mechanisms were operating on a group of arthropods that have been extinct for some 250 million years, even to the extent of suggesting comparable deviations in hormone production. Who said the fossil record cannot reveal anything about the detailed mechanisms underlying evolution?

Among living animals, experimental application of hormones to particular larval stages has shown that the different sorts of ants present in colonies, such as "workers" and "soldiers," might arise merely by slight changes in the timing of hormone production. Although looking very different, genetically they are virtually indistinguishable. Application of juvenile hormone to the last larval growth stage of one particular species of ant delayed the onset of metamorphosis and caused a longer period of growth. Instead of metamorphosing into a wimpish little worker, a fierce soldier emerged, complete with larger body size and overdeveloped jaws, ready to defend the colony to the death. A side effect of this is a change in behavior, this "overdeveloped" (i.e., peramorphic) ant being suitably aggressive. Similarly, the entire caste system of termites is based upon so-called polymorphisms that result from variations in the time that the different castes take to metamorphose. The operation of heterochronic mechanisms affecting the timing of maturation was fixed into the genetic makeup of the termites and resulted in a mix of castes that function as an integrated society. The same is probably true for many other groups of social insects.[10]

Stress and Sex in Salamanders

Not only environmental factors such as temperature, nutrients, and hormones directly influence the direction that development takes; another, more unexpected environmental factor is stress. This, too, can determine whether or not sexual maturity is triggered.

Among vertebrates, this is perhaps most elegantly portrayed by some salamanders that have evolved the capacity to switch between alternate developmental pathways, depending on environmental stimuli. This is known as *facultative heterochrony*. In salamanders, facultative heterochrony takes the form of delayed onset of metamorphosis, which produces paedomorphic individuals. One of the classic examples of such paedomorphosis occurs in the salamander known as an axolotl, or Mexican salamander. This salamander's morphological development is so reduced that it fails even to metamorphose. The result is the retention of larval characteristics, including gills, even when it is sexually mature. It also retains ancestral juvenile behavioral patterns by remaining in the juvenile aquatic environment and not metamorphosing into a land-dwelling form.

One intriguing question that has bothered a few scientists is why some populations of salamanders include paedomorphic individuals while others do not. Reid Harris of Duke University investigated this by undertaking a series of experiments that involved raising larvae of the salamander *Notophthalmus viridescens dorsalis* in tanks under different population densities. He set out to examine what effect this might have had on the generation of paedomorphic individuals. Harris set up one tank in which he placed ten larvae. In another he placed forty, giving them the same amount of food and leaving them to develop. What he found was that over the same period most of those larval salamanders kept at low population densities became paedomorphic and stayed in the water, where they bred in a juvenile form. The majority of those in high-population-density tanks metamorphosed into immature land-dwelling salamanders before becoming terrestrial adults.[11]

When you think about it in terms of what does or doesn't make evolutionary sense, there seems to be some logic to the strategies adopted by the salamanders under different population pressures. When there is lots of water and lots of food, there are lots of happy baby salamanders hatching from eggs laid in a pond. But as time passes and the baby salamanders grow bigger, the pond starts to dry up. The water level in the pond drops—there is less water and less food, but the same number of not quite so happy, slightly larger babies. Their effective population density is increasing—not by uninvited guests dropping in, but by a shrinking of resources as the drinks and nibblies run out. And the bigger they get, the more they need to eat. Under such circumstances, selection would favor those individuals that metamorphose into land-dwelling adults and disperse into a whole new supermarket of food which has suddenly been made available. For those salamanders left behind in the ever-shrinking pond, the cupboard is almost bare—if they remain as juveniles, their fate is sealed, as they will be—into a muddy tomb. But such is the way of natural selection. In this case, the flexible survive to pass on to future generations this same developmental flexibility, while the inflexible perish, and with it, their more developmentally inflexible genome.

Stress is also a major factor in the development of monstrous polymorphs in amphibians of the genus *Ambystoma*. Under certain conditions, in addition to the normal morph huge, cannibalistic monsters sometimes develop. These have enlarged heads and jaws with a correspondingly formidable feeding apparatus that allows them to feed on the smaller amphibian larvae. The reasons that the cannibals develop is thought to be because of overcrowding, the stress so induced acting on a single gene that induces increased rate of development of specific head structures. Again, associated with these changes are changes in behavior, the cannibals being more aggressive than the normal amphibians.[12]

In some ways, introducing a pressure-inducing stress component into the environment can almost be seen as compensating for an inherent laziness that has crept into the developmental program of the species. Given the good life as children, why grow up and have to cope with all of life's traumas? What did Garstang say? "Live as tadpoles, breed as tadpoles, tadpoles all together." Such latent developmental flexibility, producing paedomorphs or not at the drop of a hat (or at the drop of the water level) has been the character that has been selected for and proved to have been an immense evolutionary success story for a very long time. As I discuss in Chapter 7, the fossil record indicates that such developmental lability in salamanders has been the hallmark of the group for hundreds of millions of years.

Return of the Dog

Well, having diverged somewhat from the dogs that are the title of this chapter, and gone on what you might think has been at times a rather long-winded excursion through various elements of the animal kingdom in my attempts to demonstrate the all-pervasive influence of heterochrony within species, it would seem fitting to close this chapter with a final look at dogs. But I shall end on a somewhat cautionary note. For I do not wish you to get carried away with the idea that any and every sort of variation is possible and that, in the case of dogs, any number of breeds can theoretically be bred. On the contrary, even within species like dogs, and humans, in which certain aspects of their anatomy show a reasonable degree of variability, there are still constraints on how far it is possible to go. After all, the anatomy of both dogs and humans is constrained in certain areas, such as in the number of legs they produce, the number of eyes, the number of ears, and so on. Pere Alberch, who now works at the Natural History Museum in Madrid, has been involved for some years in much pioneering work on the whole question of the relationship between embryonic development and evolution. He has delved, in particular, into what factors determine the number of toes that amphibians can produce, and how they can experimentally be induced to change this number. He has also looked at this question in breeds of dogs to highlight the fact that there are developmental limits to the degree of variability that can be selected for, either artificially or naturally.

There is a perception among society as a whole that many scientists suffer from tunnel vision when it comes to their interaction with the world around them. I would, of course, in my own case, resolutely deny this. Just because, when a visiting colleague from England (a skilled hand with a frying pan) decided to buy some garfish, and instead of flinging them straight into the pan, we ended up sitting on the kitchen floor all evening, measuring body length, lower jaw length, and upper jaw length, doesn't mean to say that we forgot to cook the garfish afterward. After all, in our defense, I would argue that these fishes represented as fine a developmental sequence as one is likely to obtain from any fishmonger! And I have been researching heterochrony in the ancient and living worlds for the best part of twenty years, so it's not always easy to switch off such thoughts when one is out of the work environment. But the point that I am trying to make is that you shouldn't misjudge Pere Alberch's actions when he arrived home in Spain from the United States and greeted his parents, who proudly showed him their new prize possession—a Great Pyrenees dog, rather like a St. Bernard. And what was the first thing he did? He counted its toes! Well, after finishing a series of experiments on the relationship between cell number in early embryonic amphibians and the number of toes these amphibians ultimately produce, and spending a long time counting vast numbers of frog and salamander toes, I suppose that he just got used to doing a quick on-the-spot toe count of any passing animal. But there was actually a fair degree of method in Pere's apparent madness. For what the toe count did was to support a hypothesis that he had developed from his work on the amphibians—namely, that larger animals tend to have more toes than smaller animals.

Pere Alberch's toe count of his parents' Great Pyrenees revealed that the dog had six toes on each hind foot and five on the front ones. Further investigation led him to discover that most dog breeds have four toes on each hind foot and five on the front ones, the extra toe being the so-called dewclaw. Some larger breeds, however, also have this fifth toe on the hind feet as well. But in the biggest breeds, like the Great Pyrenees, St. Bernards, and Newfoundlands, a sixth toe, a double dewclaw, is often present. At the other end of the spectrum, tiny dogs, like the chihuahuas and Pekingese, lack the dewclaw on the front limbs, having only four toes all around. This means that in dogs, variability in toe number (a good example of heterochrony of parts) is constrained by the body size of the animal.[13] Extrapolating his research that he had carried out with Emily Gale, Pere Alberch explained the relationship between toe number and body size in terms of the dog's early developmental history.

When Alberch and Gale treated an early embryonic amphibian limb with a chemical that stops cell division for a period, a limb developed in the adult that was half the size of a normal limb. Not only that, but the smaller limb also has fewer toes, and the order in which the toes are lost is very specific. This means that the sequence in which they develop on a normal limb is also very specific. So assuming that large dog breeds have correspondingly larger embryos, it is not unreasonable to suggest that these larger embryos will have larger limb buds, composed of more cells, than the embryo of a small dog after the same period of development. Selection for large body size in some breeds of dogs therefore has had the effect of "dragging" along the development of extra toes. While, as Pere Alberch points out, breeders have for a long time tried to come up with all sorts of fancy explanations for the adaptive significance of the extra toes in large dogs, Alberch argues that such developmental interdependence means that certain structures need not be of particular adaptive significance.

Interestingly, cats also sometimes develop extra toes—so-called witches' cats. A cat that I had as a pet when I was a child in England possessed six toes (as expressed by its claws) on its front legs, and five on its hind legs (that's one extra on each foot). But in cats the development of extra toes is not dependent on larger body size. My cat was quite small. Maybe rather than having formed, as in dogs, by a hypermorphic extension of the period of limb development, it may occur by an acceleration in cell division and growth.

So while much of the variation that occurs in shape and size within organisms arises from tinkering with the developmental program, there are boundaries that constrain the operation of heterochrony. The success and diversity of many groups of animals arises not only, as we shall see in later chapters, from the specialized structures that they have evolved, and which equip them to inhabit a specific niche, but also from the evolution of species with flexible developmental systems. This very plasticity either allows different niches to be occupied at particular times, or else allows a single niche to be more effectively utilized.

5

Time for Sex

Sex is the queen of problems in evolutionary biology.
Perhaps no other natural phenomenon has aroused
so much interest; Certainly none has sowed as much
confusion.

G. Bell, *The Masterpiece of Nature: The Evolution and
Genetics of Sexuality*

Every female, just at the age of maturity, is more like
the young of the same species than the male is observed
to be; and if the male is deprived of his testes when
young, he retains more of the original form, and therefore more similar to the female.

J. Hunter, *Account of an Extraordinary Pheasant*

LADY TYNTE WAS, TO PUT IT MILDLY, EXTREMELY SURPRISED, AND ALSO
somewhat annoyed. For a long time she had kept peafowl, and developed an understandable bond with each of her peacocks and peahens.
What amazed her was when her favorite peahen, which had over the
years dutifully produced numerous clutches of fertile eggs, underwent a most surprising change—"she" began to look more and more
like a male, growing a spectacular train. Such dazzling tail feathers are
usually the prerogative of males. This avian transvestitism was used
by Dr. John Hunter in 1780 in the presentation of his ideas on primary and secondary sex differences. What had happened to Lady
Tynte's peahen was not an uncommon phenomenon: the peahen's
production of estrogen stopped. It is usually considered that male
secondary characteristics, such as a showy plumage in birds, are a

consequence of testosterone production. Not so, argue Ian Owens of University College, London, and Roger Short of Monash University in Australia—it is the production of estrogen by females that inhibits the growth of the highly colored feathers and gives them their drab plumage.[1] This is just one example of how the different appearance of males and females is strongly controlled by hormones. Many other examples that also involve differences in body size between the sexes can trace their origins back to variations in the timing of production of growth or sex hormones. And they can have the most extraordinary consequences . . .

A World without Sex

Believe it or not, there was once a world without sex. In the seas that swept the fledgling Earth three and one-half to four billion years ago, life evolved. It is probable that these first, primitive life forms were the simplest of cells, presumably little more than clusters of RNA (ribonucleic acid). Until the last decade, RNA had been thought to be little more than a humble messenger, transcribing strands of DNA to direct the formation of proteins. But biochemists have come to realize that in primitive life forms reproduction may well have occurred by the RNA making copies of itself, initially of its own volition, but probably later assisted by protein enzymes. This means that the first life forms on this planet were not DNA-based but RNA-based. At some stage in the early history of life, these early cells must have changed from having an RNA-based genome to a DNA genome, the single strand of not very stable RNA being replaced by a more resilient, double, helically spiraled DNA molecule—one of the first increases in complexity. Yet even with DNA-based primitive cells there was still no room for sex: on the primordial Earth, sex had no ugly head to rear up and replication took place by the much less interesting process of simple cell division.

But don't imagine that these most primitive of organisms that were swept through Earth's early seas went the way of most ancient life forms and became extinct. Not so—they (or at least their descendants) are still alive and well today. Those based on DNA persist as bacteria, while some scientists hold that many of the viruses that plague our lives today may be evolutionary hangovers from deep time, remnants of a world before even bacteria existed.[2] Today some viruses, such as those that produce the common cold and poliomy-elitis, still use RNA as their genetic material, acting for all the world as if nothing had changed around them in four billion years.

While the fossil record has so far failed to come up with any evidence of these very primitive viruslike organisms, it has, over the last decade, revealed some of the earliest, most primitive DNA-based cells. The long-playing fossil record is a repository for more than just the rotted bones of some unfortunate dinosaur that failed to look over its shoulder as it calmly drank at a waterhole. Given the right conditions for fossilization, even the simplest of cells can be fossilized. In the case of very early cells, this has allowed us a voyeuristic time-traveler's glimpse back into the ancient bacterial world.

To find this earliest evidence of life, it is necessary to trek to one of the hottest places on Earth—the Pilbara region in northwestern Australia. Much of this area is underlain by vast deposits of banded-iron formation. The source of huge quantities of iron ore, these rocks are also thought to have been the product of the activity of these primitive, photosynthetic cells. The metabolic waste product of these cells was oxygen, which converted ferrous iron dissolved in the sea into mountainous accumulations of insoluble ferric iron: the Earth literally rusted for tens of millions of years. The Pilbara region, in which these rocks occur, has been one of the most tectonically stable regions on Earth for literally billions of years, to such an extent that Roger Buick, when at the University of Western Australia, discovered a 3,500-million-year-old land surface, complete with original proto-soil. For here, at a site most inappropriately named North Pole (a fine example of sardonic Australian humor—North Pole is situated close to Marble Bar, a small town that once went for 162 consecutive days with daytime maximums over 100° F), the silica-rich sediments laid down about 3,500 million years ago remain unaltered. They persist as a testament to the fortitude of life, as well as to the resilience of the fossil record. To date, eleven different cell types have been described from these Australian rocks. Most are thin filaments, just 0.5 to 19.5 μm wide. These mere wisps of our most ancient ancestors are so similar to some modern bacteria that William Schopf of the University of California at Los Angeles, when describing them, remarked that "if they had been detected in a modern microbial community and morphology were the only criterion by which to infer biological relationships, the majority would be interpreted as oscillatoriacean cyanobacteria."[3] And as cyanobacteria they, like their modern-day descendants, would have photosynthesized and provided the Precambrian oceans and atmosphere with oxygen.

But what would the creatures inhabiting this ancient world have looked like? Dorion Sagan and Lynn Margulis have suggested that a microscopic examination of the microbial communities that inhabited these distant Precambrian seas "would have revealed flotillas of bobbing purple, blue-green, red, and yellow spheres: colonies of organisms crowding on rocks, gliding on water, or darting about with whipping tails. Shoals of bacterial cells waved with the currents, coating pebbles with brilliant hues. Bacterial spores blown by breezes showered the muddy terrain."[4]

Like other bacteria, the oscillatoriacean cyanobacteria (formerly known as blue-green algae) are prokaryotes. That is to say, they are bags of DNA, lacking the organelles, such as a nucleus and mitochondria, that are found in the eukaryotic cells of animals, plants, and fungi. The DNA in prokaryotic cells is not confined to discrete areas within the cell, such as chromosomes. Without sex, prokaryotes reproduce by doubling in size, replicating their single strand of DNA before dividing, with a copy of each set of DNA being delivered to each offspring cell.

If, as seems likely, many of the cells that inhabited these ancient seas really were cyanobacteria, then this was of great significance for the evolution of future life on Earth, because such cells would have been photosynthetic. If such a complex system of oxygen production were present in this relatively diverse 3,500-million-year-old assemblage (in fact, as Schopf suggests, twice as diverse as the average Precambrian assemblage), then life is likely to have evolved somewhat earlier than this, maybe as far back as 4,000 million years. Considering that the Earth is thought to be in the region of 4,600 million years old, it would appear that life must have gained its foothold on Earth at a very early stage in its history. As far as we can tell from the fossil record, sexless cells were the Earth's sole inhabitants for the first 1,500 million years of life's tenure on the planet. As well as inhabiting the seas, it is likely that bacterial spores—small, enclosed packets of DNA budded off parent cells—were blown onto the naked land where they probably thrived in lakes, ponds, and puddles. The world was permeated by as complex a microbial ecosystem as exists on Earth today.

The Rise of Sex

Without sex, biological evolution for the first half of Earth's history was painfully slow. Indeed, until the recent discovery in 2,100-million-year-old rocks in Michigan of some bizarre spiral structures up to 90 mm long, it had been thought that prokaryotes had ruled supreme for 2,500 million years. What these spirally coiled megascopic carbonaceous films are thought to represent are not only the remains of the earliest multicellular algae but also the earliest evidence for the existence of eukaryotic cells. And with their evolution came the possibility for sex. This different kind of cell was both larger and more complex than the prokaryotic cell. Complete with a meandering series of channels of internal membranes, one of which enshrouded the nucleus, it also carried within it other organelles, such as mitochondria: structures that provided the cell with energy from oxygen.[5]

In their pioneering work on the origin of sex and eukaryotic cells, Lynn Margulis and Dorion Sagan have championed the notion that the organelles present in eukaryotic cells, such as the nucleus, mitochondria, and chloroplasts, were originally free-living prokaryotic cells that took up symbiotic relationships with other cells. They think that by living within larger cells, smaller cells benefited from entering, and living within, a food-rich environment, as well as being afforded some degree of protection. The host cell would have prospered under this arrangement by using up the invader's waste metabolic products.[6] If Margulis and Sagan are right, such mutualistic cohabitation has led ultimately to the evolution of all life forms on Earth, other than bacteria and viruses, from turnips to truffles to trilobites.

Perhaps of almost equal significance to this symbiosis between prokaryotes is the role that heterochrony played within the eukaryotic cell replication that facilitated the secret of sex—the evolution of meiosis.[7] What meiosis achieved was the formation of eggs and sperm, and consequently the ability to exchange DNA. But exactly what is sex from the point of view of a cell? We all have a pretty good idea (at least most of the time) of whether an individual of our species is one sex or another. And it is generally pretty well known that there are basic differences between the sexes in those rather enigmatic little strands in our cells called chromosomes, so that in mammals, for instance, adding X and Y chromosomes together spells male, whereas adding two X chromosomes spells female. Eggs and sperm are essentially specialized cells that contain half the normal allotted number of chromosomes. When they come together as a primal cell with the full number of chromosomes for the species, half are derived from one individual, while the other half come from another. And therein lies the route to the production of variation between parent and offspring; between ancestor and descendant; and ultimately to evolution.

In recent years, Margulis and Sagan have promoted the work that was carried out in the late 1940s by Lynn Cleveland. She demonstrated how a delay (a postdisplacement) in the timing of onset of division of centromeres in the cell led to meiosis, and consequently to sex. This delay resulted in the number of chromosomes being halved, instead of staying constant, during cell replication. Such a delay in division by the centromeres would have resulted in only halves of each chromosome being pulled to the centrioles. When the centromeres eventually divided, paired chromosomes would have split and the cells divided, with each cell containing only one set of chromosomes.[8] Such heterochronic delay in the timing of a crucial event in cell replication lies at the very basis of the evolution of sexual reproduction and thus the evolution of plants and animals. It is likely that the first multicellular eukaryotes were algae. Fossilized impressions of spheres, joined by thin connecting tubes reminiscent of living seaweeds, have been found in rocks 1.3 billion years old, again in the Pilbara region of northwestern Australia.

Exactly what do we mean when we talk about males and females, and what role has heterochrony played in the evolution of different morphological characteristics between sexes? Most organisms produce gametes (sex cells) of different sizes. In most cases one group will produce large gametes, known as eggs, while another produces very small ones—sperm or pollen, depending on the kingdom. Basically, individuals that produce small gametes are males; those that produce large gametes are females. I shall talk at some length in Chapter 7 about how cells themselves can be shown to undergo the same sorts of heterochronic changes as entire organisms, in terms of

complexity and size, but suffice it to say here that from an assumed
state in ancestral eukaryotes, where gametes are likely to have been of
the same size, there has been differential growth between the male
and female gametes, with the female gametes having undergone a
peramorphic increase in size, relative to the male gametes. This probably reached its extreme in the extinct, flightless bird *Aepyornis*, which
lived until early historical times on Madagascar. This bird, which
weighed in somewhere in the region of a ton, produced an egg that
was up to 35 cm in length. With a volume equivalent to about
135 hen's eggs, it would have produced a sizable omelet! However,
even some sperm grow to astronomical proportions. The fruit fly
Drosophila bifurca produces gargantuan sperm that are almost 6 cm
long, or 20 times as long as the adult male's body! As they become
sexually mature, the testes of this fly increase greatly in size to occupy
more than half of the abdominal cavity.[9]

As well as differences in gamete size, many species of animals and
plants exhibit size and shape differences between sexes. This is known
as sexual dimorphism. Many of these differences, such as the ornate
plumage possessed by many male birds, or greater relative body size
in male mammals and female spiders, have been the subject of much
scientific scrutiny for over two hundred years. Charles Darwin's pioneering work *The Descent of Man and Selection in Relation to Sex*, published in 1871, was in many ways inspired by John Hunter's treatise in
1780 on primary and secondary sexual characteristics. However,
there has been little investigation into the manner in which these differences in size, shape, and behavior have arisen—most research has
centered on the adaptive significance of the differences.

I believe that heterochrony permeates sex—producing a small
male here, a larger female there—speeding up, or slowing down, the
growth of structures that allure the opposite sex or help in fighting off
members of the same sex; or tinkering with the time that each sex becomes sexually mature. While it is all very well to examine in intricate
detail the selective advantage that the tail of one male bird of paradise
may have over another, to understand the totality of the evolution of
the trait, both how it originated and what advantage it conferred on
the species' fitness, we must also look in detail at differences in the
growth and development between males and females within species
to explain how these differences arose in the first place. Indeed, the

evolutionary success of many species is a consequence of this very sexual dimorphism. Had substantive variations between the sexes never developed, then many species may well not have evolved. So, to understand fully the role of heterochrony in evolution, we must be able to recognize the part that it has played in sexual dimorphism.

One of the features of heterochrony is that some processes, in particular those affecting changes in the time that sexual maturity is attained, will result not only in differences in what the sexes look like but also whether one sex will be larger than the other at maturity. Consequently, if the period of growth that the organism spends as a juvenile is shortened, we have progenesis, and it will fail to grow as large as its ancestor. Conversely, lengthen the fast juvenile growth period and the organism is likely to grow larger, as well as change shape. Thus, if males and females of a species attain sexual maturity at different times, then there can be profound disparities in body size and shape between the sexes. And with these size and shape differences come the inevitable behavioral differences. However, as I will show, in some species there may also be fundamental differences between sexes in their growth rates, even quite early in embryonic development.

Shall It Be Male or Shall It Be Female?

Conventional wisdom has it that, in the case of mammals, male development is set into motion by specific signals sent out from the fetal testis. In males, during the first trimester of development, cells that synthesize the male sex hormone testosterone appear in the testis. In females, on the other hand, there is little cellular change until the second trimester, when ovarian follicles develop. So the earlier differentiation of the testis is in strong contrast to the much later differentiation of the ovary. Again, we are dealing with rates and timing of development. Ursula Mittwoch of Queen Mary and Westfield College, London, and her colleagues have argued that selection for early differentiation of male gonads and production of their own sex hormones, long before the female embryos start churning out their own hormones, probably arises from the development of eutherian mammals in a sea of maternal hormones. In cases where a male embryo leaves it too late to start developing testosterone, then sex-reversed XY females can form.[10] Testosterone is therefore the male's defense mechanism against being smothered by female hormones and ending up to all intents and purposes a female on the outside but a male on the inside—a confusion that most of us could well do without.

Formation of the testis actually precedes the formation of male sex organs. In females, however, formation of the ovaries occurs after female sex organs have formed. Development of an externally "intact" male is therefore dependent on the testis being formed at the right time during development in order for testosterone to be produced on cue. Differentiation into an externally apparent female is not dependent on the presence of the ovary and its hormones. So even before adolescence, there are differences in the timing and rate of growth between the sexes. However, secondary sexual characteristics and the dimorphism in body size and shape between adult humans, like other animals, is, in part, also a product of differences in the time of onset of maturity.

One of the more traumatic phases of the life of any adolescent male human is that crucial period when those silly little giggly girls that you have been sharing your class with at school almost overnight undergo a startling metamorphosis into sophisticated, elegant, attractive young women. While we humble males at this time remain as weedy, smooth-faced youths fascinated only with football, Super Nintendo games, or whatever the fad of the day is, our female classmates' interests have been catapulted in one fell swoop from Barbie dolls to an almost predatory fascination with somewhat older members of the opposite sex. The sex hormones have been launched and heterochrony is here at play, fiddling with the timing of onset of maturity between the sexes and wreaking absolute hormonal havoc. When viewed dispassionately in purely heterochronic terms, female humans are progenetic, compared with males. In other words, they attain sexual maturity earlier than males of the species.

But why should there be a difference in the time that maturity is reached? In humans, something is clearly triggering earlier production of sex hormones in females. This has the effect of severely reducing the production of growth hormones. At the cellular level, the result is the disappearance of growth plates on the bones. Bone grows from each end by means of special growth plates that are composed of cartilage. Just a few millimeters thick, these growth plates are situated at the ends of long bones, such as the femur. During growth, the cells multiply and are replaced by bone that forms from other cells, so elongating the limb. At maturity the growth plates dissipate, and bone replaces them. The ramification is that bone growth, and thus limb growth, stops.

Because girls mature, on average, about two years earlier than boys, growth plates stop functioning earlier, resulting in an overall smaller body size. For growth plates to function fully, growth hormones must be secreted by the pituitary gland, which is located at the base of the brain, and transported in the bloodstream. The growth spurt, so characteristic of puberty in humans, arises from increased sex hormone production, stimulating a greater production of growth hormone. However, interestingly, cessation of growth to a large degree arises from the continued influence of these hormones, which cause the growth plates to disappear.

So, it would seem that differences in the timing of hormone production must play a significant role in determining sexual dimorphism in humans, not only at adolescence, but also, as I have discussed, very early in development. Until recently, the established belief was that prior to differentiation of the testis in human males, the pattern of development of male and female embryos was identical and that, as a consequence, there were no apparent differences between very early male and female embryos. However, Ursula Mittwoch has recently upset the embryological applecart by showing that in different species of mammals growth rates between male and female embryos can vary, even *before* male gonads differentiate.[11] Experiments undertaken on mice and cattle have shown that very early XY (in other words, male) embryos show accelerated development, compared with XX embryos.

In humans and rats, measurable sex differences can also be detected in embryos prior to differentiation of the testis. This undermines the notion of those who have for so long argued that timing and rate variations in development between males and females are due entirely to hormonal differences between the sexes, since production of the sex hormones in this early phase of development is still some way off. Supporting evidence comes from measurements of skulls made on human fetuses. These showed that at eight to twelve menstrual weeks, growth in females was about one day behind that in males, and that at birth this had increased to about six to seven days. Thus, both before and after differentiation of the gonads, males develop at a faster rate than females. As this is set in train prior to testosterone production in males, the cause must be put down to some other factor, the most likely being the direct effect on the sex chromosomes.

Ursula Mittwoch has pointed out that even thirty years ago it was being suggested that differences in chromosomes themselves could affect the duration of cell cycles and the rate of cell differentiation. What Mittwoch has suggested is that the presence of the second X chromosome in females may have the effect of slightly reducing the rate of cell differentiation, compared with males. Why this should be so is not clear. The slightly faster growth rate engendered by Y-chromosomal genes starts, in mice at least, incredibly early in development, functioning as early as the two-cell stage.

Stag at Bay

The next time that you feel inclined to have a drink of fine Scotch malt whiskey and you reach for your bottle of Glenfiddich, look closely at the beast haughtily staring out from the label. For in your hand you are holding a drawing of what is arguably one of the most striking portrayals of sexual dimorphism—a red deer stag, the apotheosis of male sexual dimorphism in terrestrial herbivorous mammals. Such a huge red deer, resplendent in his fearsome array of antlers, strutting arrogantly among his harem of females, typifies how in many mammals males have a larger body size, as well as a dazzling array of secondary sexual features, in this case a stunningly impressive set of antlers.

In many such sexually dimorphic mammals, there is a distinct difference between males and females either in the age at which they reach sexual maturity or of relative rates of growth. For a male mammal to invest more energy in growing to a larger size than the female, growth must be of appreciable adaptive significance. One incidental benefit in mammals such as the red deer is the notably enhanced weapon growth (i.e., horns, antlers). While it can be argued that prolonging the juvenile phase in males, so producing a larger body size relative to that of the female, is advantageous, it could alternatively be argued that there is an advantage for females to mature relatively earlier than males, since this increases their chances of breeding. Most likely it is a combination of the benefits achieved from both earlier female maturation and delayed maturation in males, resulting in a relatively larger male body size and consequent improved combative capabilities, that produces the "fittest" species.

In a study of sexual dimorphism in terrestrial herbivorous mammals, Peter Jarman of the University of New England in Australia showed that this same pattern of larger males than females occurred in three-quarters of the 107 bovid species that he analyzed. Yet at birth the body weights of males and females were generally about the same. What Jarman discovered was that within bovids three types of growth patterns to maturity have evolved.[12]

In small species, such as African duikers and dik-diks, there is little, if any, sexual dimorphism, males and females attaining their adult weight at much the same time as each other, early in life, and maturing at the same time. However, in medium-sized bovids, such as impala, waterbuck, and lechwe, males reach maturity one to several years *after* the females—they are relatively hypermorphic. A consequence of this is that the males spend a longer time than the females in the rapid juvenile phase of growth, so they end up with a heavier and larger body than the females. This weight is then retained for the rest of their adult lives. The actual rate of juvenile growth is much the same in both males and females. The third bovid group includes the really large species, such as the buffalo and the American bison. Growth in the males of these animals, like the African elephant, is essentially indeterminate, whereas in the females it may or may not be. In other words, growth in males does not stop at the attainment of sexual maturity, whereas it may or may not in the females. Significantly, the females also have a slower growth rate than the males. In heterochronic terms, the females are neotenic, compared with the males, as they grow at a slower rate. The size differential between males and females therefore increases throughout the life span of the animals, and can be quite pronounced.

The intensity of development of weapons or ornaments (such as horns and antlers) varies not only between closely related species (see Chap. 6) but also between the sexes. In small bovids the presence or absence of such structures distinguishes subadults from adults, and their growth ceases at the onset of maturity. Because these structures have relatively little time in which to grow, compared with larger bovids, they are correspondingly smaller. In medium-sized mammals in which the males reach sexual maturity more than three years after the females, horns and antlers continue to grow after maturity. For example, in red deer the number of tines in the antler increases each year, through much of the adult male's life. So in bovids of this size body growth stops at maturity, but antlers and horns continue to grow. However, if antlers or horns are present in the females of such species, they cease to grow after maturity is reached.

It might be asked why in many medium-sized male bovids body growth, but not horn development, ceases at maturity. From a hormonal perspective, it would seem that the factors responsible for the curtailment of growth hormone production over much of the body at the onset of maturity result in another hormonal switch being turned on, with these hormones selectively targeting the horns or antlers. In terms of the adaptive significance of this dimorphism Jarman has argued that perhaps there is a physiological limit to body size, particularly if the species are inhabiting a seasonal, fluctuating environment. There is probably selection for cessation of growth at maturity because the number of females in a male's group will eventually be limited by external factors. Therefore, it becomes profitless to attain a larger body size. However, since courtship and mating will be very important to reproductive success, the larger the weapons the better. So selection has favored those individuals that continue antler growth. Moreover, the greater the extent of antler growth, the more likely it is that, in combination with a large body size, the individual will have reproductive success. After all, following eating or avoiding being eaten, a successful sex life comes high on most animals' agenda of life.

Pygmy Pan and the Pigtailed Macaque

The trend of larger male body size is found in many other groups of mammals. It is the case in many species of primates. As with the bovids, the larger the ape the greater the degree of size difference between males and females. Although the chimpanzee, *Pan troglodytes*, shows some distinction in size between adult males and females, there is little shape difference between the two. Body proportions and facial features are much the same in the two sexes—the body is just larger in the male. The reason for the lack of variation in shape between the sexes may be a result of the fact that the intersexual size differences are relatively small. Thus, the relatively peramorphic nature of the male is insufficient to have any appreciable effect on shape. Another possibility may be that the allometric differences are not very great in chimps. Thus, because proportionate shape changes between juveniles and adults are not very great, the disparity between the sexes will not be very great. A different situation is found in the larger *Gorilla* and *Pongo* (orangutan). In these primates the differences in overall body shape between the females and males are appreciable, particularly in shape of the face and of the braincase. Such differences have arisen as a consequence of the greater extent of size dimorphism between males and females. This is irrespective of whether or not the allometries are any different from those of *Pan*. Consequently, where the differences in maturation times between the sexes are more exaggerated, or where allometries are greater, then sexual dimorphism will be more pronounced.

Even between the two species of chimpanzee, *Pan troglodytes* and the pygmy chimp *Pan paniscus*, there are significant differences in the degree of sexual dimorphism, particularly of certain morphological features. Brian Shea of Northwestern University has pointed out how the face of the pygmy chimp grows at a slower rate than that of the common chimp.[13] The result is a relative reduction in sexual dimorphism in the dental and facial regions. In particular, there is less dimorphism in the canine teeth of the pygmy chimp. As a consequence, there seems to be lower male-male and male-female aggression in this species than in the larger common chimp. There is also increased female bonding and increased food sharing in the pygmy chimp. What we have here is a case whereby subtle differences between species in growth rates between the sexes can lead to significant behavioral differences.

In some primates, sexual dimorphism may not be simply a consequence of rate differences between the sexes, but be due to a complex cocktail of timing and rate variations over the entire growth period. Growth often does not proceed at an even rate throughout development. Teasing out the underlying processes can be quite a difficult exercise. In the pigtailed macaque, *Macaca nemestrina*, for instance, Rebecca German and her colleagues at the University of Cincinnati found that there appear to be two distinct growth spurts during development, one before the animals are one year old, the other after about two years. They also discovered that adult males are larger and heavier than adult females.[14] This arises partly from the greater growth spurt in juvenile males compared with females and partly from the earlier initiation (predisplacement) of the growth spurt. Thus, the sexual dimorphism that is manifested in adults begins to appear quite early in development. A consequence of the first phase of male growth extending for a relatively longer period than in the female is that at the end of this first phase of growth the males are already larger than females of the same age. In the second phase of growth the males also grow at a faster rate as well as for a longer period, thereby combining the heterochronic processes of acceleration and hypermorphosis, respectively. The product of these two phases of differential growth between the males and females is thus pronounced sexual dimorphism. A similar phenomenon has been reported in other primates, including long-tailed macaques, rhesus macaques, and howler monkeys.

The toque macaque (*Macaca sinica*) from Sri Lanka also shows a complex pattern of growth.[15] During the juvenile period females have two phases of growth, while males have three. Sexual dimorphism is minimal in infants and young juveniles, up to two and a half years old. However, after that, sexual dimorphism starts to become apparent. Skeletal limb growth ceases in juvenile females at about five and a half years of age but continues in males until about seven and a half years. Muscle mass shows a similar pattern, with muscle growth ceasing in females at about eight years, but in males not until twelve years. Males also undergo an adolescent growth spurt by acceleration. Thus, while much of the sexual dimorphism is a result of differences in the time that maturity is attained between the sexes, there are also relative differences in growth rates in older juveniles, which contribute significantly to final adult dimorphism. In a review of sexual dimorphism in 45 species of primates, Steven Leigh of Northwestern University found that dimorphism in many primate species arises from a combination of both bimaturism (males and females maturing at different ages) and growth rate differences.[16] Such a combination therefore leads to accentuated dimorphism, greater than can be achieved by the operation of one process alone. Both mechanisms, acceleration and hypermorphosis in the males relative to the females, culminate in the males attaining a larger body size, arguing for strong selection pressure on this feature in primates.

You should not forget that we humans too are primates, and as such should be no different from macaques and chimpanzees (at least in terms of what drives sexual dimorphism). In *Homo sapiens*, sexual dimorphism is largely (although, as we have seen, not entirely) a result of differential timing in onset of sexual maturity. Rates of post-embryonic juvenile growth in humans vary slightly between the sexes. Body size differences between the sexes arise, in general, due to the delay in onset of maturity in males. This is about two years later in males than in females, on average. Males thus continue on the faster juvenile growth track for longer than females, as in many other primates, resulting in the generally larger body size and some of the attendant characteristic morphological features. In heterochronic parlance human males are relatively peramorphic, females relatively paedomorphic. The relatively paedomorphic features retained by females include small faces, reduced cranial sinuses, decreased cranial robusticity and cresting, and relatively larger brains.

Extreme Sexual Dimorphism

Sexual dimorphism is just as common in many groups of invertebrates, ranging from the subtle to the overt in differences in body size and shape. A typical example of extreme sexual dimorphism arising from heterochrony occurs in a small bivalve called *Pseudopythina rugifera*. Females of this species have a shell that grows up to 15 mm long. The tiny, sexually mature male reaches no more than 1.25 mm in length and actually lives within the body of the female. However, these males do not remain as paedomorphic males all their lives. After they have done the deed and fertilized the host female, they leave and continue their development outside of her. But that's not all. After leaving the females they change sex, first becoming hermaphrodites, then undergoing another sex change to become females themselves, to await the arrival of their own tiny, underdeveloped male. And so the cycle of changing sex continues.

However, this pales into insignificance when compared with the extraordinary degree of dimorphism that occurs in eulimid gastropods that parasitize holothurians (sea cucumbers), such as *Entero-*

Extreme sexual dimorphism in the parasitic gastropod *Thyonicola* living within a holothurian (sea cucumber). The male is extremely reduced and is parasitic within the female, spending its life in a special receptacle where all it does is fertilize the female.

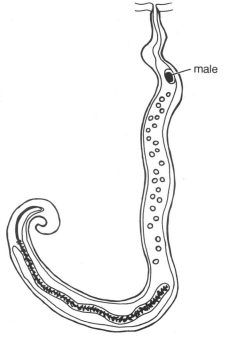

xenos and *Thyonicola*. The females of these thin, sluglike gastropods live a less than exciting existence as parasites within the intestine of a sea cucumber. But what of the males? For a long time it was thought that the species were hermaphroditic. However, a careful examination of eulimids by Jørgen Lützen of the University of Copenhagen showed that the so-called testis is not merely a male gonad but is actually an extremely paedomorphic, individual male. After having lost its larval shell, this male enters the female through a tube that connects the female to its host's esophagus. The minute, paedomorphic male then attaches itself to a special male receptacle within the female and expands into what is really little more than a testis. It needs little else. Its life is spent in connubial bliss, doing nothing else but fertilizing the female.[17]

In such species, the parasitic occurrence of the male within the female ensures an increased chance of fertilization of the eggs in a relatively unstable habitat. The lottery of tossing sperm into the ocean and hoping that it will make contact with a passing egg is forgone for a safer, more reliable method of fertilization. Thus, the strongest selection pressure ensuring optimum survival of the species is the pae-domorphic condition of the male, enabling "in-house" fertilization and high levels of fecundity. Again, had not such a large disparity in size and shape developed between the male and the female, the species may well not have evolved. By having its own resident, live-in male, the female, living its own slightly precarious life within a sea cucumber, has maximized the chances of her eggs being fertilized, allowing the species to flourish.

The size shoe is on the other foot in *Cystococcus*, an Australian scale insect that infests some species of *Eucalyptus*, producing large, woody galls. The female is extremely paedomorphic, its only appendages being genitalia and mouthparts that it uses to tap into the phloem rising through the tree. After the female has constructed the gall, she lays the first part of her clutch. These are all males. They undergo normal development, passing through a number of molts, changing into fully winged hemipterans, complete with three pairs of legs. Their only peculiar characteristic is a very elongate abdomen. About the time that the males become adults, the mother lays her second batch of eggs. These are entirely female. These are paedomorphic in that they

The adult winged male of the gall-forming scale insect *Cystococcus*, carrying his tiny paedomorphic sisters to another tree.

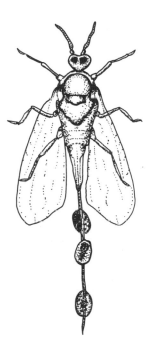

The progenetic male of the deep-sea angler fish *Photocorynus* lives attached to the head of the much larger female. The male is about 15 centimeters long, the female about 1 meter.

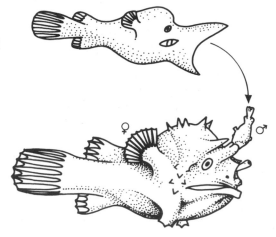

fail to produce wings. Early on they have stubby legs, but these are lost after the first molt. But before this occurs they climb on to their brothers' elongate abdomens. Once aboard, the males fly off with their female passengers to colonize another tree.[18]

Extremes of sexual dimorphism such as these are not restricted to the invertebrate world. Perhaps the most startling example among backboned animals occurs in some deep sea fish. In one group, the seadevils (family Ceratiidae), the females are very much larger than the males, up to 145 cm long. The tiny, paedomorphic males are never more than 16 cm long. In one genus, *Photocorynus*, the male attaches, parasitically, to the head of the female. In another form, *Ceratias*, the tiny male fixes itself to the underside of the female's body. Perhaps in the deep, dark depths in which these fish live the chances of successful encounters with members of the opposite sex occur only if they both are in constant contact with each other.

Swordtails and Leatherjackets

Other fishes demonstrate more subtle heterochrony-driven sexual dimorphism. One of the more classic examples of sexual dimorphism in fishes occurs in a group of freshwater fishes from Central America within the genus *Xiphophorus*, known as swordtails. As their name implies, these fishes possess a swordlike spine on the base of their tail fin. This sword is only present in males and varies in length between species. Another group of species within *Xiphophorus* are known as platyfishes. They are similar in many respects to the swordtails, but they lack the swordlike spines on their tail fins. Until recently it had been assumed that swordtails evolved from the simpler platyfishes. Such thinking is still rife in biology—that evolution always proceeds from the more simple to the more complex. Yet if one accepts that heterochrony plays a leading role in evolution, then it should come as no surprise to learn that species can evolve that are less complex in some respects than their ancestor. And they achieve this by paedomorphosis.

Support for such a scenario in swordtails comes from a recent study of the relationship between the twenty-two species of swordtails and platyfishes, based on molecular data. Specifically this involved analyzing the relative degrees of similarity of mitochondrial and nuclear DNA between the various species within *Xiphophorus*. What this study by Axel Meyer and Jean Morrissey of the State University of New York at Stony Brook and Manfred Schartl of the University of Würzburg showed was that on a number of separate occasions the morphologically less complex (with respect to their tail fins) platyfishes evolved from swordtails by a loss of the sword.[19] It has long been known that the presence of this sword in males is an attractant to females, and that females will preferentially mate with these males with longer swords. Surprisingly, even female platyfishes prefer males with swords. Attachment of artificial swords to male platyfishes results in females spending more time with these apparently now more attractive males.

The development of the swords has been shown to be under hormonal control. By applying hormonal treatment to male platyfishes, swords can be induced to grow. The "loss" of swords in platyfishes is therefore likely to have occurred by a block in the genetic cascade that would otherwise have turned on the hormone at the right developmental time to induce sword growth. The way that the female platyfishes respond to the artificially armored males is particularly interesting. If the switch that turns on sword development is just lying dormant in the platyfishes, so too is the behavioral response of the females. All this, of course, begs the question of why the swords would have been lost in the first place in some lineages if the females found them so attractive. One possible explanation is that the presence of a sword on the tail fin impedes the swimming ability of the males. This may have led to increased predation levels in some species that were encumbered by these structures and unable to outswim their predators. Such high levels of predation pressure may have outweighed the benefits that otherwise would have accrued to the males with long swords from the desire of the females to mate with these piscine King Arthurs.

In these cases of sexual dimorphism, only specific structures that play some role in sexual attraction are affected by changes in their rate of growth. In other instances among fishes, the entire shape of the body can be affected. My ichthyologist colleague at the Western Australian Museum, Barry Hutchins, has extensively studied some species of leatherjackets that are common in Australian waters. He has shown that in one species, the mosaic leatherjacket, *Eubalichthys mosaicus*, which swims in the oceans off southern Australia, there is substantial sexual dimorphism in body shape. The juveniles of the species possess a short, diamond-shaped body, with quite short dorsal and anal fins. However, during ontogeny the males undergo greater allometric shape change than the females, with the result that the males attain a longer, more slender body than the females, which retain the more squat diamond shape of the juveniles. In the males the dorsal and anal fins also become longer. The changes in body and fin shape produce attendant behavioral consequences, because since the males are so different from the females, their swimming technique is

The adult female of the blue boxfish, *Strophiurichthys robustus*, from the western Indian Ocean, is paedomorphic, compared with the male, resembling the juvenile in body shape and some color patterning.

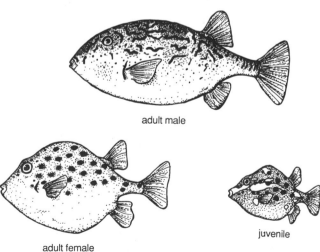

adult male

adult female

juvenile

correspondingly different. The large chinaman leatherjacket, *Nelu-setta ayraudi*, shows a similar pattern of relatively paedomorphic female, in this case the female retaining the juvenile character of a more slender body than the male. In other species of leatherjackets the females are paedomorphic relative to the males only in coloration. A striking example is the toothbrush leatherjacket, *Penicipelta vittiger*. In this fish juveniles of both sexes, along with adult females, have the same mottled brown color, whereas the adult male is a vivid blue, with a yellow tail.[20]

In another group of fishes, the boxfishes in the family Ostraciontidae, females retain both the juvenile diamond-shaped body and prominent spots that are absent on the male, such as the blue boxfish, *Strophiurichthys robustus*. Like some of the leatherjackets, the body of the males is more elongate than that of the female or juveniles. As with the leatherjackets, in some boxfishes, such as the white-barred boxfish, *Anoplocapros lenticularis*, paedomorphosis in the females is reflected only in coloration patterns. The females retain the juvenile brown coloration, whereas the males attain a striking red with white splashes.

"Come into my parlor," said the Spider . . . Having a reasonably large jungle of a garden in suburban Perth, in Western Australia, which is home to a wide variety of invertebrates, I have the chance to observe the variation in sexual dimorphism between different groups of spiders a lot more easily than the dimorphism that occurs in fishes in the seas off the Western Australian coast. During the day, few spiders are in evidence in the garden. Any large, lone wolf spider that is bold enough to venture out is more often than not grabbed by large, predatory, black and orange spider-eating wasps. Paralyzing these huge, hairy spiders, the wasps will, if need be, drag their heavy burden the entire length of the garden to wherever they have located their egg chamber, in order to lay a solitary egg in the body of the paralyzed, but still living, wolf spider—fresh food for the larva when it hatches from its egg.

But at night the spiders take over the garden. Any nighttime stroll tends to be a slightly hair-raising experience, as the chances of walking into a spider's web half a meter or more wide are quite high. The thought of a spider with a 6-cm leg span dropping down your shirt front is usually enough to deter too many nocturnal perambulations in the garden. The female spider constructs her web early in the evening and usually, when her night's passive hunting is complete, neatly rolls it up into a little ball, save for the main guy ropes which remain to be used the next night, and eats it. Not only does this turn out to be considerate to the humans who wish to walk in the garden during the day, but it is also an effective way of recycling energy, and perhaps of taking in moisture that has condensed on the web in the early morning.

As with many other invertebrates (with the notable exception of some beetles), sexual dimorphism in spiders is reflected in females often being much larger than their male counterparts. The classic story of the huge female spiders consuming their hapless mates following consummation of their passions is yet another example of intersexual heterochrony. Mating in orb spiders, where the male may be less than half the size of the female, can be a hazardous experience for the male. The male of the orb spiders that inhabit my garden begins his romantic overture by plucking at the outer strands of the female's web, like a harpist playing an enticing tune. When this has done its trick, he approaches the female very cautiously. The first contact comes when they gently touch "hands," very tentatively at first. Then in a flurry of legs they fling themselves upon each other in a passionate embrace. But is this male to end up as his love's after-sex feast? Not if he has his way. He has a most cunning escape mechanism, for he brought along with him a long escape rope. All of a sudden he releases the female, and before she can plunge her fangs into him in a less than romantic manner, he sails off in spectacular fashion like Tarzan on his long strand of web, out of her clutches.

There is great variability in the extent of dimorphism in spiders, not only between different families, but also even within single species, with the males typically being smaller. Body size dimorphism occurs more often in sedentary, web-building spiders. However, the large huntsman spider with the 10-cm leg span that sprints across my bedroom wall in the middle of the night like an arachnid Carl Lewis may well be either a male or a female. For there is little size dimorphism in active, roaming, hunting spiders. We know that the size dimorphism in orb spiders, such as *Nephila*, is the product of heterochrony, because males are known to mature relatively earlier than females.[21] This accounts for their smaller body size and less complex morphologies. The selective advantage of such differential timing in onset of maturation times is thought to be attributable to the larger female size generating higher fecundity—the bigger the female, the more eggs and the more young. With the high levels of juvenile mortality, the production of large numbers of offspring is imperative to the species' survival.

Just how high these levels of predation can be was brought home to me when I watched the newly hatched orb spiders in their special nursery web prepared by the female. Having watched one particular female spider for months, as she spun her enormous web directly outside our dining room window during dinner, my family and I felt rather protective of her young. Getting larger and larger throughout the summer, she had finally given birth to well over a hundred tiny offspring. But within a day the local ants had located them. Dragging their way along sticky webs, rather like tightrope walkers trying to negotiate their way over a wire coated in molasses, these ants, barely as large as the baby spiders, had killed and eaten most of the brood within a day. Probably just one or two survived to provide this year's garden trap. To avoid such high predation levels, selection will favor those individuals that have the fastest growth rates. The quicker they grow, the larger they get, and the harder it becomes for ants to prey upon them. Of course, this will probably take them into the size range for some other predator, before they themselves turn the tables and become predators.

In the case of orb-weaving spiders, which show high sexual dimorphism, the males benefit from a high growth rate and also from attaining sexual maturity earlier than the females. This is because, in general, being more mobile than the females, they suffer higher levels of juvenile predation. Consequently, by maturing relatively earlier, fewer male juveniles are killed. Moreover, the quicker they mature, the better the chance of mating. In the case of more mobile spiders, where the female is not sedentary, there is nothing to be gained in particular from one sex maturing any earlier than the other; hence, there is little dimorphism.

An Inordinate Fondness for Beetles

In addition to displaying an apparent inordinate fondness for beetles, God appears to also have injected them with a healthy dose of heterochrony. The effect is seen in the males of many species of beetles that develop either greatly enlarged front legs, mandibles, horns, or spectacular spear- or forklike structures on the head or prothorax. Such features are the product of positive allometry. Consequently, they are more prominently developed in larger individuals that have grown for a longer period of time. Even within a single sex, this character may be dimorphic. David Cook, when working at the University of Western Australia, showed how in the dung beetle *Onthophagus ferox* the sexual dimorphism involves, in part, the presence of horns on the male but not on the female. There are two distinctive male morphs: one with small body size and no horns, the other larger and with horns. Such male dimorphism, arising from allometric scaling of horn size, occurs in a number of other beetles, illustrating that the formation of such horns occurs as a result

Strong sexual dimorphism in the beetle *Dynastes hercules* from Ecuador. The massive cephalic horn in the male arises from an acceleration in growth of this part of the head. About half natural size.

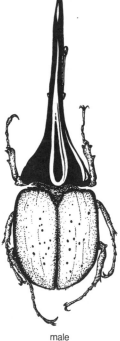

female

male

of differences in either growth rates or timing of cessation of growth, in conjunction with strong positive allometry in the structure.[22]

The size of the beetle, and consequently whether or not the male develops particularly large structures, depends, in part, on the availability of resources. In the forked fungus beetle *Bolitotherus cornutus*, for instance, Luther Brown and Larry Rockwood, working at George Mason University in Virginia, found that those individuals that as larvae feed on more nutritious fungi both emerge as adults at a larger size and attain a larger adult size. As a consequence, those males develop relatively larger horns than those feeding on less nutritious fungi. Furthermore, Brown and Lockwood found that the developmental state of the fungus can affect growth rates. Beetle larvae growing in younger fungi typically grow faster and attain a larger body size than those feeding on older, less nutritious fungi.[23] Here again we see the environment playing an active role in fashioning the appearance of the species, or one part of the species, in this case the males.

Sexual dimorphism, by definition, implies that males and females have different shapes. As such, if the argument is made that the generation of morphological variation between individuals is essentially controlled by changes in the timing and rate of development, then it becomes hardly surprising that sexual dimorphism can also be seen to be the product of heterochrony.[24] It is clear that dimorphism can be manifest at many stages of ontogeny, from the earliest embryonic stages through to the end of the juvenile phase and even beyond, in the case of one sex having indeterminate growth.

The all-pervasive influence of heterochrony in generating sexual dimorphism plainly demonstrates how fundamentally distinct body plans can be generated by heterochrony within a single gene pool. The lesson for this in the evolution of species is to reinforce the notion that distinct morphological differences between species may also be engendered with little genetic change.

6

Birds, Brachiopods, and Bushbucks: Heterochrony in Speciation

There was an old belief that in the embers

Of all things their primordial form exists,

And cunning alchemists

Could recreate the rose with all its members

From its own ashes, but without the bloom,

Without the lost perfume.

Ah me! What wonder-working occult science

Can from the ashes in our hearts once more

The rose of youth restore?

What craft of alchemy can bid defiance

To time and change, and for a single hour

Renew this phantom-flower?

H. W. Longfellow, from *Palingenesis*

STRADDLING THE EQUATOR, SOME 800 KM OUT FROM THE NORTHWEST coast of South America, is a group of small islands known as the Galápagos Islands. Unlike most Pacific islands that are (or were) swathed in green shrouds of lush, tropical vegetation, these islands are, for the most part, little more than barren, rocky protruberances—ancient volcanoes rising deep from the ocean floor. It was his experiences here, at this most unlikely location, that are thought to have laid the foundation for Charles Darwin's concept of descent by natural selection. But the selfsame animals that exemplify Darwin's ideas also graphically demonstrate how variations in the timing and rate of development are necessary in providing the raw material for natural selection to work on.

Darwin's Galápagos Finches

The generally held belief is that because the Galápagos Islands were the *Beagle*'s first port of call after leaving the South American mainland, Darwin was eager to make a large collection of the animal and plant life, about which little was known at that time. There is no doubt that he did amass a large collection of plants, marine mollusks, insects, lizards, snakes, and birds from the various islands of the Galápagos, but the reality of the situation was that Darwin was decidedly homesick at this stage of his voyage and appears to have paid little attention to much of what he was collecting. Indeed, it wasn't until a year after he had returned to England, when the ornithologist John Gould pointed out to him the great diversity of some of the animals that he had collected, that he began to draw on them for the development of his ideas on "descent by natural selection." What the specimens indicated was that each particular island had its own distinctive species. On other small islands he found that in some categories a wide variety of forms coexisted. Perhaps the best example of this was in one of the least spectacular groups of animals, the finches. Rather drab in appearance, with most males being black and most females brown, and blessed with distinctly unmelodious songs, the many species of finches differed from each other in one striking feature—in the amazing variety of sizes and shapes of their bills. As Darwin reported in *The Voyage of the Beagle* (1839): "The most curious fact is the perfect gradation in the size of the beaks in the different species of *Geospiza*, from one as large as that of a hawfinch to that of a chaffinch and . . . even to that of a warbler."

Other finches, Darwin noted, had beaks like a parrot's, another like that of a starling. To Darwin, "Seeing this gradation and diversity of structure in one small, intimately related group of birds, one might really fancy that from an original paucity of birds in this archipelago, one species had been taken and modified for different ends." In other words, evolution from an initial colonizing species.

Since Darwin's original observations, much fascinating research has been undertaken on these birds. As a result, "Darwin's finches" have taken on the role of the quintessential example of evolution.

This is, in some ways, surprising, because although certainly a beautiful example of adaptation, in *Origin of Species* Darwin doesn't actually refer to them specifically. However, since Darwin made his initial observations, a great deal of research has shown how the differences in beak shape between various species of finches reflect adaptations to feeding from a wide variety of food sources. On one island, Indefatigable Island, for instance, ten species of geospizine finches show a range of bill shapes that are used for a variety of purposes, from crushing to grasping to probing. The crushers, with their strong, stout beaks, are able to crack hard nuts and seeds; the graspers are the possessors of small, narrow beaks, capable of catching insects, while the probers have bills that are better suited to feeding from flowers or fruits. There can be little doubt that the idea of natural selection was formulated in part from Darwin's analysis of the finches, and his realization that from an early colonizer a wide range of species with different-shaped beaks had evolved, each adapted to feeding from different food sources.

The story of Darwin's finches crops up in virtually every book that has been written about evolution this century. Whole books have even been devoted to them. This is not surprising: the finches are an excellent portrayal of natural selection in operation. What has worried me for some time, though, is the manner in which a number of these books describe how evolution takes place. And this applies not only to the finches but also to descriptions of evolution in animals and plants in general, particularly in more popular books on the subject. Let us take one example: Alan Moorehead's fine account of Darwin's voyage on the *Beagle*. He writes: "On one island they [the finches] had developed strong thick beaks for cracking nuts and seeds, on another again the beak was adjusted to feeding on fruits and flowers."[1]

Now, there is no doubt that some species *do* have strong bills, and these species *do* use their bills for cracking open nuts. What I find disturbing is that simmering in the background is the implication that the birds somehow possess the bills in order to carry out a particular task—almost as though the finches deliberately evolved bills of certain shapes and sizes to fulfill a particular allotted role for each species. Now while this may to some degree be just a careless way of expressing what was happening, I think that it demonstrates the subconscious striving of many biologists to explain everything purely in adaptive terms, thus implying that the environment somehow crafts the form of animals and plants. This cart-before-the-horse approach permeates many books on evolution. In many ways it seems to smack, dare I say it, of a Lamarckian approach to evolution.

At the beginning of the nineteenth century, the French naturalist Jean-Baptiste Lamarck suggested that species could change over time as a result of the use, or nonuse, of particular parts of the body. Knowing that structures such as muscles grow with use, he assumed that an enlarged structure developed by the parent could be inherited by the offspring. Likewise, structures not used would fade away and become lost altogether. The classic example is the giraffe. Lamarck argued that these animals evolved their long necks by constantly stretching higher and higher for the more succulent leaves at the tops of trees. Any individual stretching its neck just that little bit higher would pass this higher neck on to its offspring. While Lamarck's ideas were discredited by Darwin's concept of natural selection, the way some people, particularly in the popular media, describe evolution, one would think that this sort of thinking still persists. Species, it is argued, evolve a particular structure so that they can function in a certain way—"The lesser spotted cordwangler evolved a long bill so that it could feed on the tubular flowers of the fardang tree" or some such. This sloppy way of describing evolution does little to help us in our quest to understand how evolution really works.

When I talk about the cart-before-the-horse approach, what I am driving at is that the role of intrinsic factors, most notably, changes in rates and timing of growth of structures, such as the finches' bills, must be taken into account more fully, rather than being an after-thought to the main emphasis of what the structure is used for. Any attempt to explain evolution only in terms of the selective advantage of one shape over another is telling only half of the story. Variations in the shape of the bill within a single population must have occurred *first* in order for one particular trait to be more successful than another. That means we must first look at how such variation was generated before explaining of what use it would be. Let us take a theoretical example of a group of founding finches, which long ago landed on one of the Galápagos Islands. No other finches or finchlike birds had been there before them. Now let us suppose, as seems likely, that these finches had quite small bills that were suitable for feeding upon small seeds, and that similar seeds existed on this particular island. To our pioneering finches the island would have seemed like a veritable Utopia: a sumptuous, uneaten feast littering the trees and the ground.

Seeds, of course, vary greatly in size themselves (also by differences in seed growth rates or length of growth—see Chap. 9). Let us suppose that our pioneering finches would have fed on seeds within a particular size range. Seeds outside this range, both smaller and larger, would have been ignored. Now, within this population of finches, not every bird would have had a bill of exactly the same size. During their ontogeny, both within the egg and as hatchlings, the rate of growth of the bill would vary slightly between individual birds. Similarly, some birds would have perhaps developed for either a slightly longer or a slightly shorter time in the egg or during the hatchling period. This is likely to have resulted in birds appearing in the population with either marginally larger or smaller bills. Moreover, the upper and lower bills themselves might have grown with slightly different relative growth rates and produced individuals in which the differential in size and shape between them was more pronounced. A certain range of bill sizes and shapes would lie in a range of optimum fitness. In other words, the bill functioned as efficiently as possible in its task of enabling the bird to feed from the small seeds within the species' size range.

Not only would the size and shape of the bills have varied, so too would the seeds upon which the finches fed. It may have been that no birds within the population were able to feed from the larger seeds, which had correspondingly thicker shells and were harder to break open, because the bills were of insufficient size and strength. However, over time a small number of individuals are likely to have evolved within the population that had larger, stronger bills. Such birds would be able to feed on the much more plentiful supply of larger seeds that would otherwise remain uneaten. In this regard they could be considered to be fitter, inasmuch as they had evolved a larger, stronger bill that happened to be able to cope with a more bountiful food supply. From such individuals that passed on this trait, birds with even bigger bills could evolve. Let us suppose that there existed a second species of tree which had appreciably larger seeds, and which hitherto the finches had been unable to feed from. If the trend toward the evolution within the population of some individuals with larger bills continued, then a time would be reached when individuals would appear that were able to feed from the larger seeds from the second tree. If this tree was essentially geographically isolated from the other species of tree, then the establishment of a population of rather extreme variants of the founder species would soon become geographically isolated themselves from the founder population. As a consequence they would become genetically isolated. Over succeeding generations the accumulation of slight, but substantial, genetic change between the two populations might mean that if individuals from each population came together at a later stage, they would be unable to interbreed, to produce viable offspring.

Here is how the evolution of a new species might occur, taking into account both intrinsic (changes in bill growth) and extrinsic (suitable availability of resources) factors. In this scenario it is necessary for the morphological changes to be generated first to allow descendant birds to feed on a food source that is slightly different from that of their ancestors. But the morphological changes that ensued did not occur as a result of large-scale genetic changes. They would have occurred as a corollary of maybe a slight delay, or perhaps an earlier onset of production of a particular growth hormone responsible for initiating and controlling bill growth. Alternatively, a slightly increased quantity of growth hormone may have been produced that targeted the bill, causing forms with slightly larger bills to evolve. The substantial genetic differences that exist between sister species are likely to have built up *after* the critical morphological changes and the reproductive isolation that arose between the two populations. Consequently, the heterochronic changes occurring within the population that would have been preferentially selected would depend on the environmental conditions pertaining at the time. If these extrinsic factors in some way changed, then a different set of heterochronic morphotypes might have been selected.

The evolution of the long, slender bills of the nectar feeders, in contrast to the stout bills of the seed eaters, is a consequence of different parts of the birds' bills growing at different rates. Growth in a structure such as a bill will not occur equally in all directions. If it did so, then the bird would end up with a spherical bill! Rather, growth is greater along certain vectors than others. Variations in growth rates along different vectors are responsible for the evolution of distinct bill shapes. One can only but presume that at some stage in the evolutionary history of the Galápagos finches, extreme variants of one population were ill adapted to feeding from seeds but found a new food source in flowers. Perhaps at some stage in their evolution birds existed with an intermediate bill shape, able to feed from both food sources. Certainly the honeyeaters that flit through my garden in Western Australia have a somewhat catholic diet. In addition to using their elongate bills to probe deep into the clusters of flowers that make up the banksia and bottlebrush inflorescences, they are equally adept at either catching moths on the wing or snapping the large female garden orb spiders from under the eaves of my house, where they hide during the day. The wide range of species of finches adapted to feeding from a wide variety of food sources do so on the Galápagos Islands not because they have evolved a particular bill shape in order to feed from this abundant food supply. They do so because heterochronic change resulted in the production of the raw material that allowed natural selection to pick those shapes and sizes from the population that were able to function most effectively within a particular set of environmental parameters. The coming together of particular extrinsic factors, like the shape and size of a seed, with the intrinsic factor, that is, the amount of growth and development of the bill able to feed most efficiently on the seed, has been described as the attainment of the species' adaptive peak. Those individuals within a population that are the fittest in evolutionary terms are the ones that have climbed highest up the adaptive peak. But change one parameter, either extrinsic or intrinsic, and they soon come tumbling down off the peak into the valley of extinction below.

Two scientists have devoted much of their working lives to the study of the Galápagos finches: Peter and Rosemary Grant of Princeton University. Among the many fine studies that Peter Grant has carried out on these birds is one, in conjunction with Lisle Gibbs of the University of Michigan, that demonstrates how the scenario that I have outlined for the evolution of different species can be seen to be still operating today. What these findings show is how easily the best adaptive peak climbers of today can quickly be pushed off the peak when environmental conditions change.[2] One of the largest of the finches, which possesses a correspondingly large, strong bill, is *Geospiza fortis*. The presence of El Niño conditions in the Pacific Ocean

When El Niño is active, the influence on the Galápagos Islands can be seen in variations in bill size in the finch *Geospiza fortis*. When rainfall is high and soft food is plentiful, selection favors smaller beaks. During droughts, when only hard seeds are available, larger, peramorphic bills are preferentially selected.

has a strong influence on the type of bill that is selected for in this species. El Niño events occur when ocean temperatures in the equatorial eastern Pacific increase substantially. The cyclical waxing and waning of oceanic temperatures causes a corresponding cyclicity in rainfall. The effect of El Niño on the Galápagos finches is such that when there are drought conditions, forms of *Geospiza fortis* with larger adult body size are preferentially selected. Accompanying larger body size is the possession of a relatively larger bill, arising from the allometric nature of the growth of this structure. Such birds are able to feed from large, hard seeds that otherwise would be ignored when smaller, more readily obtainable seeds are present under more amenable conditions. During these periods of low rainfall, the amount of seed set will be reduced, causing increased pressure on the limited resources. As part of the natural range of variation within the species, some individual adult birds will be larger than others and will have correspondingly larger bills. It is these individuals that will thrive more during these drought periods. When the pendulum switches the other way during the El Niño cycle and there are periods of higher rainfall, seed biomass will be high. Under these conditions, selection favors smaller birds with smaller bills, which can feed on the more abundant, softer seeds. It is not too difficult to see how this same interplay between the variation of body form engendered by heterochrony within a species and changing environmental conditions can operate as the basic mechanism driving the evolution of new species.

I stressed earlier how the focus of many evolutionary studies has been on natural selection, rather than on the intrinsic aspects controlling changing shape and size. An exception to this in the case of the Galápagos finches was a study of changes in the rate and timing of growth of particular morphological features in Galápagos finches by Peter Boag, when at Oxford University. He investigated rates of growth and changing allometries during ontogeny in a number of characters in two species of finch, *Geospiza fortis* and *Geospiza scandens:* weight, wing length, tarsus length (as a measure of leg length), and bill length, depth, and width. Boag made a number of important findings. He found that the larger species of *Geospiza* grow faster than the smaller ones. Weight, wing length, and tarsus length increase relatively rapidly during ontogeny; bill characters, on the other hand, grow more slowly. The distinctive morphological variations that occur both within and between species, especially in these characters, arise from a combination of factors: differences in size of the finches at hatching, variability in growth rates, and the duration of the growth—all aspects of heterochrony. Growth rates of different characters will vary during ontogeny, such that early in ontogeny a structure may grow fast, then slow down later, whereas another takes over and increases at a greater rate. In this way these features are in accord with the principle that resources are allocated at any particular time to the growth of the components with the currently highest functional priority.[3]

In these species of *Geospiza* the tarsus growth is completed first, then the weight, followed by wing length. Full development of the beak takes some two to three times longer than this. These prioritized stages of development of the body parts are appropriate to the needs of the developing birds. Tarsus growth helps the bird break out of the egg and establish itself in the best possible spot in the nest. The weight gain provides body fat and insulation to allow the bird to become thermally independent, rather than relying on the parent for its heat requirements. Feathers then follow, followed finally by a fully functioning bill. The development of the very large, gaping mouth within the egg provides an important signaling function. Indeed, it has been suggested that a paedomorphic extension of light colored beaks and a fleshy gape may prolong the birds' parental dependency period, and also advertise to territorially inclined adults that they are just immature birds. I shall talk more about these extensions of particular periods of ontogenetic development in later chapters (in particular Chaps. 8 and 12).

The last point I wish to emphasize, which I have alluded to earlier, and which will be of relevance to the ensuing discussion of the importance of heterochrony in the development of other birds, is how the various elements of the bill are affected by differential growth. I have talked about the importance of variations in bill growth in determining bill shape and hence function. But as Boag has pointed out, and as I stressed earlier, allometric differences between different vectors within the bill contribute to the wide variety of shapes that occur within the Galápagos finches and, as I show below, even more spectacularly in other groups of birds. Of the two species that Boag investigated, *G. scandens*, with its relatively longer bill, has, not really surprisingly, a stronger bill length allometry. In other words, compared with increase in overall size of the bill, the length of the bill increases more in *G. scandens* than in *G. fortis*.

One of the more intriguing aspects of evolution is how completely unrelated groups of plants or animals manage to evolve particular structures that not only look alike but also function alike. Birds have wings that they use for flight; so do bats. Fish have fins with which they propel themselves through the water; so do whales and dolphins. Mice have legs to support their body weight and provide locomotion; so do cockroaches. Wings, fins, and legs change shape and size through ontogeny, and they each have specific functions that are selected for: the capacity for flight, swimming, or terrestrial locomotion. This phenomenon is known as *convergent evolution*. The examples I have just mentioned are among the most obvious; but there are countless examples in the animal and plant kingdoms, often of a more subtle nature. Such convergences in the appearance of structures have the potential to cause havoc for systematists trying to unravel taxonomic relationships between species. When is a character shared by two species evolved from a common ancestor, and when is it the result of convergent evolution?

Another widely held concept in evolution is that evolutionary change is random. I would disagree. There are many factors that act

Heterochrony can target specific structures, like bills in these Hawaiian honeycreepers. Some, like the akialoa, may evolve much longer upper and lower bills by peramorphosis, whereas in others, like the nukupuu, only one bill may undergo this increased growth. The consequence in terms of feeding behavior can be tremendous.

to constrain evolution and to channel it down a certain number of broad, and sometimes not so broad, pathways. Again, I consider that the dual effects of intrinsic changes arising from heterochrony, combined with the nature of the agents of selection, limit the number of options available for evolutionary change. The ontogenetic development of an organism is a built-in constraining factor in the first place. For instance mammals, unlike insects, do not produce six appendages. We can manage only four. Likewise we produce just one head. I will shortly take you on a journey to the arid regions of Australia to demonstrate how the coincidence of ontogenetic pathways and environmental gradients that play a major role in selecting particular morphotypes is one of the more potent factors in determining the direction that evolution takes. What I would stress here is that in groups such as birds, for instance, we can see parallel developmental changes affecting the growth of bills in many different groups.

The niche occupied by finches on the Galápagos Islands is occupied by honeycreepers (Psittirostrinae) on the Hawaiian Islands (although many became extinct in historical times). While the bills of the finches change from short and pointed to deep, but still short, the bills of the honeycreepers present a wild variety of shapes and sizes. From the conservative short, narrow beaks of many species of *Loxops* have evolved groups, such as species of *Psittirostra*, which, as their name implies, have parrot-shaped beaks: short but very deep, and strong enough to crack open the toughest seeds. Here there has presumably been an acceleration in the depth vector of the bill. The other extreme has been taken by species of *Hemignathus*, which have undergone no acceleration in depth, but instead a great acceleration in length of the bill, producing curving bills more than twice the length of the head that allow these species to feed from nectar that nestles at the bottom of long, tubular flowers. In species such as *Hemignathus lucidus* the upper and lower bills have their own individual ontogenetic trajectories, with the upper bill showing stronger allometric growth than the lower bill.

As niches similar to those in the Galápagos and Hawaiian islands exist in Australia, that is, tough seeds and long, tubular flowers, and because changes in basic growth rates affect the bills of Australian honeyeaters in a similar way, we see similar shapes evolving, although not to such an extreme degree as in the honeycreepers. In the Australian honeyeaters, increased growth in the depth of the bill has resulted in the evolution of one group, known as friarbirds, which are fruit eaters. The most conservative, probably ancestral honeyeater, the striped honeyeater, is purely insectivorous; but others, like the wattlebirds, as I have indicated, feed on both insects and nectar. Species with the longest, narrowest beaks are entirely nectar feeders. So here we see the coincidence of similar patterns of developmental change, and similar selective agencies, that is, the presence of broadly similar ranges of food types, producing convergent evolution in the one very significant aspect of birds' morphologies. One distinction between the honeyeaters and the Galápagos finches and honeycreepers is the absence in the Australian birds of forms with short, deep bills, that feed on seeds and nuts. This is probably because the seed- and nut-eating niche is occupied by parrots and finches. It is possible that this tells us something about when these groups evolved in Australia. The implication would be that parrots and finches evolved in Australia before the honeyeaters.

Brachiopods after the Storm

It may be stretching the imagination a little, but it can be argued that the effect of El Niño has played another, somewhat different, role in our understanding of the part heterochrony plays in evolution—not in the Galápagos Islands but in the semi-arid region of northwest Australia. After tracking north for over 1200 km from the southwest tip of the continent, the coastline of Western Australia angles abruptly to the northeast. If we zoom in on this angle of coastline, we see that there is a notch—Exmouth Gulf. If you were to step ashore here and travel south through the prickly acacia scrub, you would soon notice the ground rising to a low range of hills—the Giralia Range. As you continue into the Range, the impact of turbulent cyclones that swing across the coastline here in March or April each year, more frequently when the El Niño giant is resting, shows itself in the thousands of gullies that scar this sparsely vegetated landscape. Indeed, in the very week that I was writing this, Cyclone Bobbie was ripping across the Giralia Range with winds in excess of 200 kph, and dumping more than 400 mm of rain in less than a day. A godsend not only for the pastoralist, but also for the paleontologist.

The first geological reconnaissance of the Range was undertaken in the early 1950s. This work quickly established that the region was very fossiliferous. All the best fossil localities that these geologists found were in the central and southern parts of the Range. When I first went to the area in 1979, it took me a while to figure out why these excellent field geologists of the Bureau of Mineral Resources in Canberra never discovered the fossiliferous gullies in the northern part of the range that I and my colleagues at the Western Australian Museum were to work on for the following decade and a half. The simple reason was that they just didn't exist then. It wasn't until I stood at the boundary fence between a sheep station on part of the Range that had run a lot of sheep over the last thirty years, and one that had not, that I realized what had happened to open up this fossiliferous gold mine. Prolonged grazing on the sparse, acacia-dominated vegetation by sheep has removed much of the plant cover. Any seedlings that spring up after one of the intermittent falls of rain are quickly consumed, leaving bare, exposed ground. As the annual cyclones wheel in from the Indian Ocean to the north and dump hundreds of millimeters of rain (the record is over 700 mm in one day) on this soft, deeply weathered limestone, ever-lengthening scars cut deeper and deeper into the Earth. As they do so, fossils come pouring out, exposing evidence of one of the most dramatic periods in the history of the Earth—the time that the dinosaurs died.

In other parts of the world, rocks of this age, formed at the Cretaceous/Tertiary boundary (known colloquially to geologists as the K/T boundary) some 66 million years ago, contain either the remains of dinosaur bones or, in places like southern France, countless millions of fragments of dinosaur eggs. Here, in Western Australia, the effect is no less startling, but because the region was probably more than 50 m under the sea 66 million years ago, dinosaur remains are few and far between. In fact, to date we have recovered just two dinosaur bones and one pterosaur (flying reptile) bone from this area. But ammonites, those extinct, coiled-shelled relatives of squid and octopus, akin to the living pearly nautilus, occur in an abundance unmatched anywhere else in the world in rocks of this age. A layer of calcareous marl 2 m thick marks the last resting place of the animals that swam in the seas of the Cretaceous world. In addition to nearly thirty species of ammonites, more than twice that number of other mollusks (marine clams and snails) have been described. After a cyclone has ripped through the area and uncovered another prehistoric treasure trove, so many fossils litter the ground that we have been forced to buy a wheelbarrow to load the fossils in for the collections of the Western Australian Museum.[4]

Whether or not you subscribe to the view that an enormous asteroid or meteor rammed into the Earth 66 million years ago, wreaking absolute global atmospheric havoc, the so-called K/T boundary mass extinction event was of epic proportions. Exposed on the side of the gullies in the Giralia Range is the actual boundary: 2 m of very fossiliferous, soft limestone, overlain by a thin bed of greensand. This rock takes its name from the mineral glauconite, a phosphatic mineral that forms only in the marine environment, where cold waters well up from the dark, icy, abyssal depths. This suggests that the greensand was formed in a cold ocean. The underlying richly fossiliferous Cretaceous limestone, on the other hand, tells of a time when the oceans were much warmer. In a geological instant, oceanic temperatures must have dropped sharply. With this came a wholesale change in the nature of the marine fauna, the like of which the planet had not seen in over 180 million years, since an even more catastrophic mass extinction event at the Permian-Triassic boundary wiped out 96 percent of marine species. Gone were the dinosaurs and flying reptiles on land, the mosasaurs (giant marine reptiles) and ammonites in the sea. But what was these groups' nemesis breathed life into other groups. For these cold waters, the first of the Cenozoic Era, were not bereft of life. A few forms had weathered the K/T boundary storm — one in particular, a little brachiopod named *Tegulorhynchia boongeroodaensis.*

If you take a walk along an ocean beach today, the chances are that among the flotsam and jetsam washed up will be lots of bivalve (clam) shells. If you had had the chance to walk along a similar beach, say, 250 million years ago, you would have been hard-pushed to find many bivalves in the debris cast up by the waves. What you would have been likely to find, though, was their Paleozoic equivalent — brachiopod shells. Looking superficially like bivalves in possessing a pair of hinged, (usually) calcium carbonate shells, the animals that lurked within would have looked quite different — not like your tasty component of moules marinière, but a feathery, probably very chewy, spiral structure and tough leathery muscles. The story of the successful rise of the bivalves to prominence through the Mesozoic Era is paralleled by the demise of the brachiopods. Although they are still with us today, only rarely are they seen, many now inhabiting very deep water. By the time of the K/T boundary, the only groups of brachiopods that really still flourished were the terebratulids and the rhynchonellids. The former are, by and large, smooth-shelled forms, the latter ribbed. Yet for all the overall evolutionary superiority of bivalves over brachiopods through much of the Mesozoic and into the Cenozoic, the thin greensands that formed after the K/T calamity shook the world show us that, apart from a few small oysters, brachiopods were the dominant bivalved shelly invertebrates.

As brachiopods go, the species known as *Tegulorhynchia boongeroodaensis,* which dominates the Giralia greensand, is one of the more beautiful. Covered by a radiating spray of up to eighty fine ribs, this species is the earliest member of an evolutionary lineage that stretched from these 60-or-so-million-year-old rocks through to the present day.[5] This lineage is particularly interesting in that it demonstrates a range of evolutionary phenomena which, when integrated, exemplify many of the principles that I have discussed in earlier chapters: in particular, the role of intrinsic changes in developmental rate and timing and selection by the environment of certain morphological characteristics. Furthermore, it illustrates how the fossil record can be used in evolutionary studies. Paleontologists have long languished under an inferiority complex when they have come to talk about fossils and evolution ever since Darwin, in the *Origin of Species*, looked upon the geological record as

> a history of the world imperfectly kept, and written in a dialect; of this history we possess the last volume alone, relating only to two or three countries. Of this volume, only here and there a short chapter has been preserved; and of each page, only here and there a few lines. Each word of the slowly-changing language, more or less different in the successive chapters, may represent the forms of life, which are entombed in our consecutive formations, and which falsely appear to us to have been abruptly introduced.

Drummed into us by generations of biologists that the fossil record is so incomplete that its usefulness in evolutionary studies was nigh on zero, many paleontologists in the past were reticent about extrapolating too much from their data. Yet since the 1850s the tens of millions of fossils collected, from the tops of the Himalayas down to oceanic depths from boreholes, has provided, for many fossil groups, a very detailed knowledge of their distribution through space and time. Obviously the fossil record is better for some groups of animals than for others. No one would disagree that the fossil record of butterflies is, to say the least, a trifle spotty. But many marine invertebrates with hard parts have left a fossil record of their attendance on Earth so complete that subtle changes within or between species can be tracked over long periods of time. Indeed, for some invertebrates that live quietly, burrowed in sediment deep in the oceans, we know as much, if not more, about their fossils than we do about the living forms.

The other sword of Damocles that has hung over the head of many a paleontologist who has tried to interpret evolutionary patterns within his or her particular group is the question of evolutionary trends. This has long been an emotive issue in paleontology. There was a school of thought, dating back to the last century, which contended that there was a built-in directionality in evolution, with a progression from more simple to more complex patterns of morphology and behavior. The term for this directionality was *orthogenesis*, and its proponents seemed almost to believe that this internal mechanism was driving evolution to bigger and better things. Some even argued that the "power" of orthogenesis was so strong that evolutionary trends could even drag lineages inexorably down dead ends and produce characters that ultimately would lead to their demise. An example of this was the extinct Irish "elk," *Megaloceros giganteus* (which was actually neither an elk nor exclusively Irish, but is more appropriately known as the giant deer). This deer possessed such huge antlers, up to 3.5 m across, that they were once thought to have ultimately led to the demise of the species. However, as I discuss in Chapter 9, recent research has revealed that nothing could be farther from the truth. Another classic "example" is the evolution of the horse—the simple trend of reduction in number of digits in the foot to produce a hoof, and of increase in complexity of the teeth to allow a change from a browsing to a grazing method of feeding. But again, as I shall discuss in Chapter 9, interpretation of horse evolution as such a simple, unidirectional pattern in a single lineage is extremely misleading.

Although the notion of orthogenesis has been completely discredited, its effect has been to make the concept of evolutionary trends noticeably unpopular, as well as making many paleontologists very wary of discussing the topic. Indeed, evolutionary trends are often envisaged by other scientists as being little more than figments of paleontologists' trend-oriented imaginations. I would argue that examples like the *Tegulorhynchia* lineage (which I will finally get around to discussing shortly) demonstrate not only a nice example of heterochrony in evolution, but also that the fossil record can play an important part in helping us interpret evolution at the specific level, not merely at the level of overall general patterns. Moreover, such lineages, and others like them that will crop up in various chapters of this book, show that evolutionary trends at the species level have been, and no doubt still are, real phenomena, in which heterochrony plays a crucial, and central, role.

After that slight digression, let me return to brachiopods, for genera like *Tegulorhynchia* are quite useful in such studies. Indeed, often where they do occur as fossils, they occur in large numbers, allowing not only general phenotypic variation within the population to be established but also ontogenetic variation. For many groups of invertebrates that live offshore in relatively cryptic habitats, we know more about their morphology from fossils than we do from their living counterparts. The earliest member of the *Tegulorhynchia* lineage, *Tegulorhynchia boongeroodaensis*, in addition to having as adults two shells covered by eighty or so fine ribs, was almost spherical in shape. Running down the center of one valve was a deep valley, with the result that if you pick up a shell and look at it end on where the two valves meet, this junction between the two valves forms a prominent arch. The other characteristic feature of *T. boongeroodaensis* is its possession of a tiny hole at the pointed end of one shell, the pedicle valve, close to where the two valves are hinged. It was from this hole (the foramen) that during its life a fleshy stalk known as the pedicle emerged from the shell and attached the shell to some suitable stable substrate, like another shell or a small piece of rock.

By studying the growth series that we have discovered in the Giralia Range over many field seasons of collecting, we are able to see that each of these features—ribs, shell shape, and size of foramen—varied substantially during its ontogeny. The ribs increased in number: the tiniest shells a couple of millimeters long have barely more than a dozen. As the shell increased in size, so the number of ribs increased. The broad valley that runs down the otherwise convex shell deepened as the shell got larger; and the shell grew as it inflated, the youngest shells being very slim in comparison with their stout parents. Lastly, the foramen started life as a relatively large hole, through which presumably a correspondingly large pedicle came. Yet as the shell grew, so little plates either side of the hole increased in size, causing the hole effectively to shrink in size. This means that the tiny shells would have had a proportionately larger pedicle than the adults. This makes sense, for even though the deeper water environment in which the species lived would have been relatively quiet, gentle currents are far more likely to have dislodged a juvenile than an adult shell. Futhermore, establishing a firm grip early in life is of paramount importance to an ambitious brachiopod.

So being armed with all this ontogenetic information about our ancient *Tegulorhynchia*, we can start hunting around for other species of the genus occurring in younger rocks. Although younger rocks of Eocene age are present in the Giralia Range, no species of *Tegulorhynchia* have been found. This is, in some ways, not surprising. For as we will see from the way that the lineage evolved, unless the environment of deposition changed in the area, then we need to look elsewhere for species that have evolved different morphologies to cope with living in different environments, probably geographically separated from the ancestral species. In the case of the next oldest *Tegulorhynchia*, we have, in fact, to travel to New Zealand, where conditions appropriate to this descendant species have been preserved and exposed to present-day agents of erosion. This species is of interest for another reason. Although it is described as *Tegulorhynchia squamosa*, brachiopod expert Daphne Lee, of the University of Otago, considers that the species is indistinguishable from a living species, *Tegulorhynchia doederlini*. If this is so, here we have evidence of a species that has existed for about 40 million years. The largest, most developed adults of this species, which first appeared in the Late Eocene, have characters like immature *Tegulorhynchia boongeroodaensis*: fewer ribs (about sixty); a slightly less portly shell; a less strongly developed median valley, and a slightly larger foramen. In other words, all these are paedomorphic characters. We know from living specimens of *Tegulorhynchia doederlini* that this species inhabits deep, quiet waters.

If we hunt for the next species in the lineage, we find it in sedimentary rocks in Victoria, where a species called *Tegulorhynchia coelata* occurs. It continues the trend established in the earlier species: still fewer ribs, narrower shell, weaker valley, and larger foramen. This trend is continued into the Early Miocene (about 20 million years ago) in South Australia. The result of continuing this trend even further is to produce species with adult shells so dissimilar from *Tegulorhynchia* that they are placed in a separate genus, *Notosaria*. A fossil and a living species from New Zealand are quite close to each other and are so paedomorphic in appearance that they resemble just a greatly enlarged tiny juvenile of *Tegulorhynchia boongeroodaensis*. The shells of *Notosaria* have only about twenty ribs, are not very convex, and have a relatively huge foramen and a stout, thick, fleshy pedicle. When the environment in which the living *Notosaria nigricans* is seen, the reason for such a morphology becomes readily apparent—they live in a very high energy environment, attached to rocks in the intertidal zone where they are buffeted by the waves.

This same pattern of a sequence of more and more paedomorphic species was one that I came across so many times in studies I carried out on lineages of trilobites and echinoids, as well as brachiopods like *Tegulorhynchia*, that I called them *paedomorphoclines*. The corresponding pattern also occurs, with sequences of more and more "overdeveloped" peramorphic species. These I have called *peramorphoclines*. I believe that herein lies the key to how heterochrony can generate evolutionary trends. Such "intrinsically" generated trends, rising from disturbances in the species' ontogenetic programs, can only arise if the paedomorphic or peramorphic variants produce a morphology that makes them able to be better adapted to a different, but adjacent environment. Many different environments form part of continuously variable gradients, such as deep to shallow water, or coarse to fine-grained sediment, or low altitude to high altitude, or low to high temperatures. It is the coming together of the intrinsic morphological gradients with the extrinsic environmental gradients that provides a channel along which evolutionary trends can develop.

In the case of the *Tegulorhynchia-Notosaria* lineage the environmental gradient is clearly deep to shallow water, since we know the environments inhabited by the end members. We can look at the changing morphological characters and explain what adaptive significance, if any, there was to the characters that were selected for over time. In this brachiopod lineage, the most critical features were the size of the pedicle and the shape of the shell. As I have already pointed out, since the juvenile was relatively more unstable than the adult, a relatively larger pedicle by which it could keep a tight grip on a hard substrate was a matter of life or death for the small, juvenile shells. The foramen size (and thus pedicle size) was in fact controlled by the rate of growth of the small plates on either side of the hole. In the *Tegulorhynchia boongeroodaensis*, these grew rapidly during ontogeny and caused the restriction in foramen size, and so pedicle size. A probable neotenic reduction in growth rate of these plates through successive species along the paedomorphocline would have resulted in younger species attaining a progressively larger adult pedicle. This has allowed these younger species to inhabit shallower water where hydrodynamic activity is higher. With the evolution of *Notosaria* came the evolution of a brachiopod with a large enough pedicle to grip firmly enough on to rocks in the highly energetic intertidal zone, without being dislodged and smashed to pieces.

A further paedomorphic character that was beneficial in shallow water was the low-convexity shells. This implies that the shells possess a relatively smaller lophophore, the spirally, feathery structure that characterizes brachiopods. This structure is used in both feeding and respiration. Such a shell form contrasts with the highly convex shell present in ancestral adults due to greater growth of the lophophore, a prerequisite for inhabiting a low hydrodynamic regime in

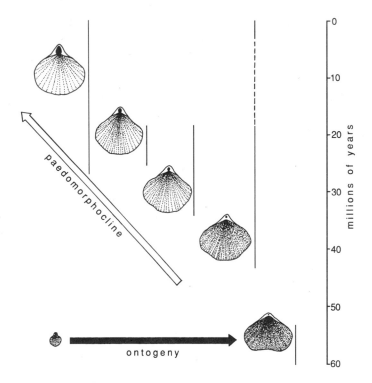

Over nearly 60 million years, brachiopods of the genera *Tegulorhynchia* and *Notosaria* have become "more" paedomorphic. Such evolutionary trends are called "paedomorphoclines." In this lineage, the retention of ever-younger features by descendant adults has resulted in the colonization of shallower water environments.

deep water. With low current activity, the possession of a larger lophophore, a structure capable of generating local current flow and increased water flow into the shell, was a prerequisite for the ancestral species living in deeper water. So, while selection might have primarily focused on pedicle size (without which attachment would not have been possible), the incidental reduction in lophophore size was a useful added adaptive feature that has allowed the living species of *Notosaria* to reach the end of the environmental gradient in the intertidal zone.

The question needs to be asked, in such examples, What actually drives evolution in this direction? Why weren't the earliest *Tegulorhynchia* content with their lot in the deep water? Was it merely a passive spread as the lineage suffered a Star Trek syndrome and "boldly evolved where no *Tegulorhynchia* had evolved before"; or was something pushing the evolutionary trend? The key to discovering whether the trend was "passive" or more actively driven lies in the pattern of the species themselves. In the case of these brachiopods we have what I call a "staggered-fork" pattern. In this pattern, many of the prongs of the fork may end at the same time, but their starts were staggered. This means that there was overlap of species: ancestor and descendant lived at the same time, for perhaps many millions of years. This is not too surprising, given that the species were evolving into different environments (shallow water), so there should have been little contact between them, therefore little or no competition, except at the extremities of the species' ranges. What drives this sort of lineage is really a passive drift in one direction. The species cannot evolve back along their gradient, because descendant forms that were peramorphic, rather than paedomorphic, would be outcompeted by the existing ancestral species. In such a case there would be nowhere else to evolve but up.

The second pattern, which I will discuss at length in Chapter 11, is one that perplexed me for many years. It is one in which a paedomorphocline or peramorphocline existed, but when the descendant species evolved, the ancestral species became extinct, even though, as with the case of the brachiopods, evolution of one species to another resulted in the occupation of a distinct niche where there would again be little or no competition between species. The key to explaining that pattern I discovered in some little sea urchins that used to live off the southern coast of Australia 20 or so million years ago. But more of that in Chapter 11. What examples like the *Tegulorhynchia-Notosaria* lineage show is that, first, evolutionary trends are real phenomena; second, heterochrony is a potent mechanism for the evolution of species; third, without heterochrony there would be nothing to select and no evolution. On the other hand, without the right environmental conditions in the right place at the right time, there would also be no evolution.

The Evolutionary Spiral

But what of the great diversity of life that we see around us today, the product of three and one-half billion years of evolution? How can we explain that in similar terms to the evolution of *Tegulorhynchia boongeroodaensis?* Is there any reason why the same processes operating on the developmental program of this brachiopod, and the pressures imposed by the exigencies of the environment, should not equally well have acted in a similar manner to fashion the fauna and flora that exist at this particular instant of geological time that we happen to call now? And can similar paedomorphoclines or peramorphoclines be identified across a range of closely related living species? If you think of the "staggered fork" pattern of evolutionary trends that I alluded to earlier, then a slice in time across the fork will reveal those species that coexist in time, though perhaps not in space. Ancestors and descendants may coexist in time. If relationships between them are paedomorphic or peramorphic, then patterns of evolution and directionality of evolution of the various species can be inferred.

To close this chapter, let us examine some animals that could hardly be further removed from marine brachiopods of tens of millions of years ago—living, breathing horned bovids, from southern and eastern Africa. For the same essential pattern of evolution that we see in *Tegulorhynchia* also permeated the evolution of certain bovids, reaffirming the crucial importance of changing rates and timing of growth trajectories in determining how species evolve. In 1990 I had the good fortune to visit a number of game parks in Zimbabwe. Never one to pass up the opportunity to look for living examples of heterochrony, it soon occurred to me that many of the differences that I was seeing between species of bovids, particularly in body size and the shape and size of their horns, were indicating the omnipresent hand of heterochrony. In impala, for instance, the young males go through stages when their horn development is like that of firstly an adult steinbuck, very short and straight; then like an adult of a reedbuck—straight, but a little longer; before finally attaining their elegant adult spiral shape. This similarity of impala growth stages to adults of other bovids is a consequence of the direct ontogenetic correlation between larger body size and more complex horn development seen across many bovid lineages. It does not necessarily indicate any direct evolutionary relationship between impala, steinbuck, and reedbuck.

In many bovids the horns spiral as they grow larger. The longer their period of growth, or the faster they grow, the more they spiral as they enlarge. This spiraling arises from differential growth rates within the horn itself, with one side growing at a faster rate than the other. In one group, the tragelaphines, the spirals arise from a faster rate of growth of the thin main sheath growing against slower growing points on the circumference of the horn.[6] With such a built-in differential, the rate and length of growth will have a profound effect on the shape and size of the horn. In bovids, as in other cervids, this is of particular importance to the males because of the importance of their horns, both in combat and in establishing dominance within the herds. To determine the impact of heterochrony on bovid development, it was necessary to investigate a group of closely related species. This group I soon saw staring out at me quietly from stands of msasa (*Brachystegia*) and mnondo (*Julbernardia*) trees in the shape of the kudu.

Of all the animals that I saw in the Zimbabwe bush, the most spectacular was the greater kudu, *Tragelaphus strepsiceros*. A magnificent beast, the males, which weigh up to a third of a ton and are almost two and one-half meters in length, sport a spectacular set of spiraling horns. Each horn in mature adults goes through two to three spirals and may reach up to one and one-half meters in length. These horns lengthen during male ontogeny, being short and straight in young males but rapidly lengthening and increasing the number of spirals through the attainment of maturity and beyond. The genus *Tragelaphus* contains a suite of morphologically very diverse species, from the greater kudu down to the little bushbuck (*T. scriptus*). The males

of this species weigh no more than 125 kg, and are barely over a meter and a half in length. Between these two extremes are four other
species, the sitatunga (*T. spekei*), nyala (*T. angasi*), mountain nyala (*T.
buxtoni*), and lesser kudu (*T. imberbis*), which vary in body size and
weight and extent of development of the horns, as well as inhabiting
different habitats. The shape and size of all these species of *Tragelaphus* are intimately tied to differences in the rates and extent of development, as I shall briefly show. But how can we decode which way
the evolution occurred? Was it from a little bushbuck-like ancestor,
through larger and larger species, to the greater kudu by peramorphosis? Or was it, like *Tegulorhynchia*, by a progressive paedomorphic reduction in size and complexity, leading ultimately to the bushbuck? Or are these each the current members of independent but
closely related lineages? To determine the polarity of evolution, we
can turn to the fossil record to help solve the directionality of the
evolution of these animals.

The earliest evidence from the fossil record of members of the
genus *Tragelaphus* is found in the emergent lineages of bovids that
sprang up in east Africa during the Miocene, as the savanna increased
in area, and habitats diversified as the world slowly began its down-
ward climatic spiral into the last great Ice Age. The earliest bovids are
thought to have been small, cryptic, inconspicuous animals, according to Jonathan Kingdon of Oxford University. Like many bovid lineages the earliest species appear to have been small, later forms
becoming larger. From an ecological standpoint, the existence of
greater food sources as new habitats open up results in selection for
larger body size, as well as greater numbers of individuals. The earliest *Tragelaphus* of which we are aware appears in fossil deposits of
Late Miocene age, formed about 7 million years ago. The species is
similar in size and morphology to the Sitatunga. This species is a little
larger than the bushbuck, and with correspondingly longer horns
that rather than being straight, as in the bushbuck, have a slight twist.
The larger tragelaphines, the 100-kg nyala, 200-kg mountain nyala,
and 315-kg greater kudu, appeared later geologically. With larger
body size, horn development is greater. So in the nyala the horn has
one open spiral; the mountain nyala, two spirals; and the greater
kudu, two to three spirals. What we see here is a peramorphocline,
with evolution to larger body size and accompanying increased spiral-

Horn development in a sequence of living bovids of the genus *Tragelaphus* from Africa shows an increase in degree of spiraling as body size increases, from the small bushbuck (125 kg) through the sitatunga and nyala to the greater kudu, males of which can grow to 315 kg (*top right*).

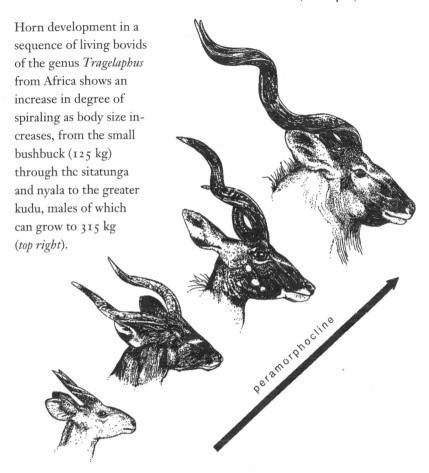

ing of the horns. However, one species, the lesser kudu, *T. imberbis*, goes a little against this neat trend, being only about the size of a sitatunga but carrying horns like the greater kudu. This probably indicates that there was a localized increase in rate of horn growth compared with other *Tragelaphus* species.

Other features also follow this general trend of peramorphosis. One in particular is the patterning on the face. In addition to their horns, one of the other attractive features of these animals is their body patterning: white slashes on the flanks, neck, and face. The development of the facial patterning not only accords with the general peramorphic trend but also provides support for the suggestion that the lesser kudu underwent a localized increased rate of growth of the horns. In the little bushbuck, the only facial patterning is a single spot on the cheek. In the larger sitatunga, an incomplete chevron-shaped bar appears beween the eyes, and two to three spots are present on each cheek. In the next largest species, the nyala, the white chevron does not appear in the females but is present as a complete bar in juvenile and adult males. In the larger mountain nyala and greater kudu, the white chevron is complete in both adult males and females. In these larger species, two to three spots occur on the cheeks, except for the greater kudu, which has three. As to the lesser kudu, its facial ornamentation is like the similar-sized sitatunga, with an incomplete chevron and only two spots.

While we can place these living species, apart from the lesser kudu, on a peramorphocline of increasing size and horn development, with the greater kudu we are not seeing the acme of the evolution of this lineage. That has already passed. While the greater kudu appeared in the Early Pleistocene, in the Late Pleistocene an even larger form existed. Described as *T. strepsiceros grandis*, this form cohabited with many other giant forms. It was at the end of the lineage of increasing body and horn size and was a beast that was perhaps up to 400 kg in weight and stood 1.6 m at the shoulder, with horns even longer and more spectacular than those of the greater kudu. It has been suggested that such a large form must have had an extensive foraging area and occupied a greater range of habitats than the greater kudu does today.

Each of the living tragelaphine species occupies a different habitat. The smallest, the bushbuck, which retains the most ancestral features, is the most widespread, occurring from forest fringes across to the dry Sahelian zones. The sitatunga has a very specialized aquatic habitat, possessing long, splayed hooves that allow it to move freely in swampy terrain. The larger nyala, like the lesser kudu, is a savanna species, whereas the largest species, the mountain nyala and greater kudu, occur principally in forests and hilly country.

From the dry thickets of the Galápagos, to the cool seas off Australia tens of millions of years ago, to the high veldt of southern Africa, the fashioning of the multitude of species that inhabit this planet has been controlled by the same intrinsic factor: changes in the rate and timing of ontogenetic development. What these and countless other examples demonstrate is how evolution of one species to

another occurs, more often than not, by utilizing preexisting morphologies—by shuffling the "amount" of growth in different parts of
the organism. By what I like to describe as species "sliding up and
down their ontogenies," new morphologies can evolve, allowing new
niches to be occupied, often tracking along environmental gradients.
The coming together of ontogenetic and environmental gradients
can then result in evolutionary trends. And what research over the
last decade has shown is that evolution can proceed in both directions—from the anatomically more complex to the less complex, by
paedomorphosis; or from the less complex to the more complex, by
peramorphosis. Evolution will go in whatever direction it can.

7

The Peter Pan Syndrome

There's a kind of glandular equilibrium.
Then a mutation comes along and knocks it
sideways. You get a new equilibrium that
happens to retard the development rate.
You grow up; but you do it so slowly that
you're dead before you've stopped being
like your great-great-grandfather's foetus.

Aldous Huxley, *After Many a Summer
Dies the Swan*

He was extraordinarily agitated now. "I don't
want ever to be a man," he said with passion.
"I want always to be a little boy and to have
fun. So I ran away to Kensington Gardens and
lived a long long time . . ."

J. M. Barrie, *Peter Pan*

STANDING ON TOP OF A SMALL BRONZE PEDESTAL IN KENSINGTON GAR-
dens in London is Sir George Frampton's sculpture of Peter Pan.
Three others were made, and one, a little surprisingly, stands not far
from where I work at the Western Australian Museum in some beautiful gardens called Queen's Park, in Perth. Surrounded by a coterie
of fairies, rabbits, mice, snails, and squirrels, J. M. Barrie's Peter Pan
epitomizes the quest for eternal youth—forever young, never growing up, and living, as Peter says, for "a long, long time . . ."

Shaped Like a Child

I have talked at length in some of the preceding chapters about the Peter Pans of the natural world: of the males that live their entire lives as tiny juveniles within their mates, yet are still capable of reproducing; of the amphibians that, in Walter Garstang's words, "think aquatic life is bliss, terrestrial a curse" and that "live as tadpoles, breed as tadpoles, tadpoles altogether"; of the increasingly juvenile appearance of the brachiopods *Tegulorhynchia* and *Notosaria* as the lineage evolved inexorably into new, shallow-water habitats. Yet the last decade and a half has seen us trying to drag ourselves away from these quests for eternal youth; not in ourselves, maybe, but in the organisms that have inhabited this planet for over half a billion years. If Stephen J. Gould is right (see Chap. 2), then the rigorous analysis of any group of organisms should demonstrate that paedomorphosis is as equally common as peramorphosis. Having, to a large degree, thrown off the shackles of the scientific cringe that permeated evolutionary thought for much of the twentieth century, and accepted that it is now politico-scientifically correct to explain many evolutionary changes by increases in developmental complexity (or, as Haeckel would have it, by recapitulation), then we should see a world where some lineages increased in complexity by peramorphosis, whereas others simplified by paedomorphosis. But is the real world as even-handed as all that? Could there not be some groups that show paedomorphosis more than others? Can people like Gavin de Beer, who espoused the cause of paedomorphosis so much, have been so blind? How can we even start to address such fundamental questions of evolutionary theory? And if, indeed, some groups show a paedomorphic tendency, can we unravel the underlying factors that might have been producing so many Peter Pans for so long?

Gavin de Beer's interpretation of the role of heterochrony in evolution consisted essentially of demolishing any hint of recapitulation while fervently espousing the countercause of paedomorphosis. De Beer believed that paedomorphosis was a particularly potent force in the evolution of major evolutionary novelties. For instance, he saw its effects in the evolution of vertebrates, humans, insects, flightless birds, flying fishes, and various crustacean groups.[1] Each, in its own way, trapped former juvenile traits into descendant adults and, as a consequence, opened up major new evolutionary pathways. While in a few groups, such as vertebrates and insects, it may still be hard to explain their evolution in any way other than by paedomorphosis, in some, like flightless birds, it played only a part (see Chap. 8). What de Beer failed to do was to undertake any sort of impartial analysis of heterochrony operating *within* particular groups. If a few groups owe their existence to paedomorphosis, are there any others in which paedomorphosis dominates? Or is Gould correct in arguing for equal opportunity for both paedomorphosis and peramorphosis?

In 1986 I was pushed into trying to answer at least one of these questions, when my colleague Mike McKinney suggested that I present a paper at a conference he was organizing in San Antonio, Texas, on the question of the frequency of different kinds of heterochrony over the last half-billion years. "Thanks a million, Mike," was, I seem to recall, my first, rather less than enthusiastic, response. How on Earth, I asked myself, was it possible to start trying to test such a proposal? The difficulties were huge. First, it can only be based on the scientific literature, and the number of examples of any kind of heterochrony described are embarrassingly small. Then there is the question of bias. There is no doubt that even now many scientists feel more comfortable describing examples of paedomorphosis than they do peramorphosis—in many ways they are easier to spot. Then there is the tyranny of the past. To many biologists the thought of explaining evolution in terms of something that even remotely smacks of recapitulation is enough to send them rushing for the smelling salts. The very idea is redolent of the moldy smell of the nineteenth century.

The only way I thought I could come up with some sort of feel for natural, rather than anthropomorphic, biases in the fossil record was to scan the literature from the time roughly since Gould's *Ontogeny and Phylogeny* was published in 1977. So, having a six-week sojourn in Cambridge in England, I hid myself away in the Sedgwick Library and scoured the literature. I focused primarily on the pure paleontological journals published between 1976 and 1985, adding in any other examples that I knew had been published elsewhere, or that I came across in other scientific journals. I found about one hundred direct references to heterochrony in the fossil record, and another forty cases in which examples of evolution had been recorded that were clearly heterochrony, although they were not described in such terms.[2]

The results of this review were most enlightening, in two ways. First, it revealed that in a few groups, such as gastropods and dinosaurs, for instance, virtually no research had been undertaken on the topic (this has, to a very small degree, been rectified in the last decade). Second, not all groups showed equal examples of paedomorphosis and peramorphosis. I was expecting that, in most cases, I would find a bias toward paedomorphosis, because of the ingrained prejudices of the past. But one group in particular, the amphibians, showed such an overwhelming bias toward paedomorphosis that it seemed likely there must be some underlying biological foundation for its dominance, especially as many examples of paedomorphosis, such as the axolotl, had been described from the living fauna. In trilobites, a group of arthropods that are now long extinct, the frequency of paedomorphosis seemed to have varied over time. But it is the amphibians, which seem to have favored paedomorphosis over hundreds of millions of years, that provide us with an opportunity to investigate possible causes for this paedomorphic bias.

Take a Walk on the Wet Side

I introduced you to the pale, languid axolotl in Chapter 4, and to how salamanders like this have the ability to switch to alternative life history pathways, one of becoming aquatic Peter Pans, the other to one of metamorphosis and a life sequestered on land. The fossil record indicates that this ability has been with amphibians in one form or another for at least three hundred million years. Of the twenty examples of heterochrony in fossil amphibians that I was able to find, seventeen were examples of paedomorphosis. Not only were these isolated examples, but a number of entire families of Permian and Triassic temnospondyls, living between 200 and 250 million years ago, have been shown to be paedomorphic. The structures that give away their paedomorphic nature are found in both the head and the limbs. The most common expression of paedomorphosis in salamanders is the retention of gills. This can be seen in fossils by the presence of gill ossicles. These are bony structures that supported the gills. Another feature that characterizes these amphibian Peter Pans is the shape of their head. Many of the early groups, such as capitosaurids, which show few paedomorphic tendencies, underwent pronounced changes in skull shape as they grew up. In the young juveniles the head was particularly short and broad, with relatively very large orbits for the eyes. As the amphibians grew, their heads became relatively much narrower and very long. In paedomorphic forms, the animal is much larger than a corresponding ancestral juvenile but retains its underdeveloped head shape, along with features like gill ossicles.

If the small size of some of the Paleozoic amphibians like *Branchiosaurus*, *Doleserpeton*, *Microbrachis*, and *Tersomius* was a product of truncated juvenile growth, then it is possible that such paedomorphs evolved by progenesis. But here we enter into one of the more contentious areas of studies of heterochrony—interpreting particular mechanisms from fossil material. Unless we have actual growth rate data (which can be found in some fossils that show growth rings in shell or bones), then we have to rely on size as a proxy for time. While this is often a good guide (for example, domestic cats reach sexual maturity and stop growing at an earlier age and smaller size than lions), it is not an invariate rule. In the case of amphibians, Paleozoic forms, like *Microbrachis*, were relatively much smaller than the nonpaedomorphic forms that metamorphosed into land-dwelling adults, never exceeding 300 mm in length. They possessed characteristic paedomorphic structures such as lateral line sulci, gill ossicles, unossified carpels and tarsels, small limbs, and an elongate trunk. If we assume growth rates similar to that of its ancestor, then its smaller size can be attributed to precocious onset of sexual maturity, truncating growth. On the other hand, it could be argued that such forms just grew at an overall slower rate, but for as long as the ancestor. In this case, the process would be neoteny, rather than progenesis. New research on bone development in another group of vertebrates, the dinosaurs, is revealing that growth rates can actually be estimated from bone microstructure (see Chap. 9).[3]

Other fossil paedomorphic amphibians may be of a size similar to nonpaedomorphic forms. Such size differences suggest that two quite different processes can act to produce paedomorphosis; a similar duality in paedomorphic processes has also been documented in living salamanders. If, as seems likely, the small fossil species arose by progenesis (in other words, they became sexually mature earlier than

Paedomorphosis has been rampant in amphibian evolution for a long time. The snakelike lepospondyl *Phlegethonia* (*right*) lived about 300 million years ago during the Carboniferous period. Another lepospondyl, the 60-centimeter-long *Diplocaulus*, was more Napoleon-like, with a very broad skull. Such broad skulls are characteristic of many juvenile Paleozoic amphibians, but the great shieldlike extensions reflect a localized increase in growth in this region. Drawings by Jill Ruse.

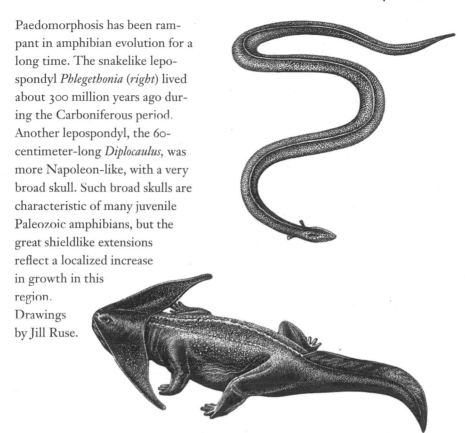

normal, so failing to achieve as large a body size, and juvenile features were frozen into these precocious adults), then paedomorphic salamanders that grew as large as nonpaedomorphic forms did so in some other way. This would have been by neoteny, as the rate of development of morphological features was reduced. So when maturity was reached metamorphosis had not been achieved, freezing them for eternity like Peter Pan.

There is little doubt that extrinsic factors also play a crucial role in the evolution of so many paedomorphs. After all, one of the characteristic features of amphibians is their biphasic life history, with juveniles inhabiting an aquatic environment and adults a terrestrial one. If a food source was abundant in the water and there was no adverse selection pressure against staying aquatic, then there was strong selection pressure favoring those amphibians able to breed in a juvenile

The presence of gill arches in adults of the
Late Paleozoic labyrinthodont *Branchio-*
saurus indicates that these amphibians
were paedomorphic. Drawing by Jill Ruse.

form in a juvenile environment. It is particularly significant that not
only the characteristic juvenile morphology is retained, but also the
juvenile habitat and behavioral patterns (see Chap. 10). Be that as it
may, there is increasing evidence to suggest that intrinsic factors may
have been equally important in this dominance of paedomorphosis in
some groups of amphibians. In living amphibians, paedomorphosis
has most often been recognized in plethodontid salamanders. In
many lineages, such as those of *Ambystoma, Aneides, Bolitoglossa, Batra-*
choseps, and *Notophthalmus*, selection for paedomorphosis has targeted
a range of morphological traits. Particularly significant are the retention of a short tail, elongate body shape, and fully webbed hands and
feet; reduction, and even loss, of some of the bones in the hands and
feet; and overall reduction in the degree of ossification in the skull.
There is also paedomorphic reduction and simplification of the lateral line and the auditory and visual systems. Simplification of the
auditory system is reflected by a reduction in the development of
structures of the inner and middle ear. The visual system shows a reduction in the number of neurons and the extent of morphological
differentiation of central processing areas in the eye.

As with the fossil amphibians, the operation of progenesis is suggested by the small size of some amphibian genera, combined with a
contracted juvenile phase of development. A variety of ecological factors have played a part in the selection of paedomorphs, most notably
the effect of fluctuating environmental conditions, such as temperature, food, population density, and habitat. As well as these global

heterochronic changes, there are many cases of localized heterochronic changes affecting only certain structures, many of which
(though by no means all) are paedomorphic.

If, as seems likely, the dominance of paedomorphosis over peramorphosis in amphibian evolution is a real phenomenon, then we are
faced with the question of what are the fundamental intrinsic factors
that have made paedomorphic structures the main target of selection
for hundreds of millions of years. Research in recent years has indicated that this "ease of going paedomorphic" may well be driven from
within, by cellular constraints. This research has focused not on the
rate of cell growth, as might be expected, but on cell size, in particular the impact on the size and shape of the organism by changes in the
size of the genome. It is well known that much of the DNA present in
cells is of no particular use to the functioning of the cell: this is the
so-called junk DNA. The question is, to what extent does the remaining, or secondary DNA, determine what an organism looks like?
More specifically, what is the significance of differences in the size of
the genome to rates of morphological development, and therefore to
heterochrony?

The *genome size* has been defined as the mass of DNA in an un-
replicated genome—in other words, how much DNA is packed into
the cell. This also goes under the name of the *C-value*. It is known
that C-values within species are remarkably constant. However, between species there can be huge variations. Despite this, the variation
in size of the genome bears no relationship to the complexity of the
organism. Nevertheless, there does appear to be a direct correlation
between genome size and the rate of development. Four different
hypotheses have been proposed to explain evolution in genome size
and the effect that these changes may have on rates of growth and
development.

The first, the *Junk DNA hypothesis*, suggests that a good deal of the
genomic DNA is essentially redundant, comprising functionally inert
sequences that are merely reflecting ancient, nonfunctional relicts of
gene duplication. So the bigger the cell, the greater the amount of
this evolutionary baggage. The second hypothesis, *Selfish DNA*, proposes that the only function of secondary DNA is just to increase the
amount of DNA in the genome, and thus increase its size. In this sce-

nario there would be a progressive increase in genome size over vast tracks of time, stopping only when the genome gets so big that it imposes no survival advantage to the organism whatsoever, because it restricts rates of growth and development.

The third hypothesis, known as the *Nucleotypic DNA*, implies that rates of development may actually be determined by the size of the nuclear genome, in an inverse manner. In other words, the more DNA that is present, the slower the rate of growth of the cell, and the slower the rate of growth of the organism as a whole. The last hypothesis, *Skeletal DNA*, considers genome size as being largely an adaptive feature, since it maintains a metabolically favorable ratio between cell volume and nuclear volume. It proposes a generalized inverse relationship between evolutionary changes in size of the genome and developmental rates.

One of the leading proponents of this idea, Tim Cavalier-Smith of the University of London, wrote that

> natural selection acts powerfully on organisms to determine their cell size and developmental rates (which are inversely related). The mean cell volume of an organism is the result of an evolutionary compromise between conflicting selection for large cell size and rapid developmental rates: the particular compromise reached for a particular species will depend on its ecological niche and organismic properties. Since larger cells require larger nuclei, selection for a particular cell volume will secondarily select for a corresponding nuclear volume, producing a close correlation between cell and nuclear volumes in different organisms. I suggest that the basic nucleotypic function of DNA is to act as a nucleoskeleton which determines the nuclear volume; small C-values are therefore required by small cells with small nuclei, and large C-values by large cells needing large nuclei. The DNA C-value of an organism is therefore simply the secondary result of selection for a given nuclear volume, which in turn is the secondary result of the evolutionary compromise between selection for cell size and for developmental rates.[4]

On the basis of available evidence, a direct correlation can be shown to exist between genome size and a number of certain characters, including cell volume and weight, length of cell cycle, duration of meiosis, pollen maturation time in plants, minimum generation

time in flowering plants, and in amphibians the time taken for em-
bryogenesis from fertilization to the hatching tadpole stage. It has
been suggested that large C-values have often been the principal targets of selection. Yet this may not always be the situation. For instance, among organisms with eukaryotic cells, mammals, birds, and
reptiles show remarkably little variation in their genome size. It is just
two- to fourfold, rather than ten- to one hundredfold, which is the
case in other groups of multicellular animals and plants. So if one is to
argue that selection focused on genome and cell size, it favored small
cell size predominantly in amniotes (the group containing turtles,
lizards, crocodiles, birds, mammals, and their fossil relatives).

Of particular interest to us here is the observation that there is frequently a strong inverse correlation between cell size and developmental rates. In other words, when cells are larger, with a larger genome, developmental rates are less than when the cells are smaller.
This indicates that cell size may well be an important factor in determining particular styles of heterochrony, especially when, as in the
case of amphibians, there appears to be a strong bias toward one type
of heterochrony, in this case paedomorphosis. Yet despite this, the
relationship between genome size and rate of embryonic development is not always a simple correlation, as for instance in frogs.
Stanley Sessions and Alan Larson of the University of California at
Berkeley have suggested that slow developmental rates in these animals are the result of the possession of large genomes. However, in
those frogs that have small genomes, there is a wide range of developmental rates.[5]

Paedomorphic salamanders are known to have exceptionally large
genomes and a correspondingly high DNA content.[6] While most
vertebrates have a genome size of about 1–3 picograms (pg) of DNA,
in salamanders the range is 14–83. Furthermore, because of the positive correlation that exists between genome size and cell volume it has
been possible to measure cell volume, and so interpret genome size,
in some fossil material. This has allowed assessment of long-term
phylogenetic changes in cell volume. In both amphibians and flowering plants there is, as I have indicated, an inverse relationship between
developmental rates and cell volume. This has been confirmed in
amphibians by Heather Horner and Herbert Macgregor at the Uni-

versity of Leicester in England who measured the C-value in eighteen species of amphibians and showed that species with the largest genomes take much longer (up to twenty-four times longer in some cases) to reach a state of development comparable with species with smaller C-values. They consider that over time there will be a tendency for genome sizes to increase unless checked by natural selection.[7]

In a similar vein, Sessions and Larson compared the C-value of twenty-seven species of plethodontid salamanders with their rates of development. They did this by analyzing the regeneration rates of limbs that had been removed. In the species that they analyzed, the C-values ranged between 13.7 and 76.2, almost a sixfold difference. Variation within any given species was low. Moreover, regeneration rates similarly showed about a sixfold difference between the slowest and the fastest. When the two are plotted against one another, they show an inverse relationship between C-value and regeneration rate; as predicted, those with the larger genome volumes had slower regeneration rates. When viewed in terms of heterochrony, it could therefore be argued that paedomorphic organisms that have evolved by neoteny, a reduced rate of growth, did so because their slower developmental rates were a product of longer cell cycles.

Animals that have large cells often have lower respiratory rates. It is also possible that selection for lower respiration rate may have contributed to favoring the evolution of larger cells and consequently large genomes. To take this one step further, it could be argued that the paedomorphic nature of animals with large cells is little more than a by-product of selection for lower respiratory rates, rather than direct selection for larger cell size. These sluggish animals with low respiration rates are likely to spend a good deal of their time under starvation conditions. Consequently, it is possible that the more economical energy metabolism made possible by larger cells may be sufficient to favor selection of large cells. It is known that amphibians that possess relatively large cells are generally pretty torpid animals. Moreover, such forms tend to be paedomorphic and do not metamorphose to air-breathing adults. Cavalier-Smith has suggested that paedomorphic plethodontid salamanders, which often have unusually high cell volumes, lack lungs, and respire through their skins, may have lost their lungs in order to grow larger cells.

Cell Size Evolution in Lungfishes

The fishes that possess the largest cell volume are lungfishes, which undoubtedly do utilize lungs to respire. Like amphibians, lungfish are able to survive low levels of oxygen, supporting the idea that selection was for low metabolic levels, favoring large cell size. Whereas amphibians utilize gills and lungs at different stages of their life histories, lungfishes utilize both methods at any stage. In addition to gills, lungfishes possess lungs that are derived from outpockets of the gut, which itself was derived from the ancestral primitive osteichthyan swim-bladder. The possession of an elongated roof of the mouth allows the lungfishes to gulp air.

The lungfish was first discovered living in Brazil in 1836 and described by the Curator of Reptiles at the Imperial Museum in Vienna, Leopold Fitzinger, as "undoubtably a reptile," on account of the remnant lung and unusual nostrils placed near the upper lip. Although the British anatomist Richard Owen was also perplexed by the first lungfish he saw, which had been collected from the Gambia River in west Africa, Owen considered it to be a fish. Anyone who has seen one of the three remaining species of lungfishes feeding, either *Lepidosiren* from Brazil, *Neoceratodus* from Queensland in Australia, or *Protopterus* from Africa, is unlikely to forget the sight. Food is taken into the mouth and chewed up using the crushing tooth plates that also characterize the fish. This masticated mess is then spat out as a long, pulpy tube, before being chewed over and over again until it is soft enough to swallow.

The earliest lungfish, *Uranolophus*, from 390-million-year-old Early Devonian rocks in North America, possessed similar tooth plates to the living forms and, like other early lungfishes from the Devonian Period, is thought to have fed in a similar manner.[8] The fossil record indicates that Devonian lungfishes were diverse and abundant. What is more, they were exclusively marine. However, since the Carboniferous Period they have inhabited freshwater environments only, and by the Permian Period, some 250 million years ago, some had developed the ability to estivate in mud. During these periods, through to the Mesozoic, lungfishes underwent pronounced morphological change. Not only were these changes driven by paedomorphic mechanisms similar to those that have so influenced amphibian evolution, but there is evidence from the fossil record that the same underlying reason, increased cell size, may also have been the cause.

Throughout lungfish evolution there were some pronounced trends toward more paedomorphic morphology. These include the loss of a heterocercal tail, fusion of the two equal-sized dorsal fins, general reduction of fin rays, loss of cosmine from the skeleton, change in scale shape from rhombic to round, and overall reduced ossification.[9] By measuring a range of lungfishes over their 390-million-year evolutionary history it is possible to show that there was a steady increase in osteocyte (bone cell) volume during lungfish evolution.[10] It is possible that, as with amphibians, this increase could have arisen from selection for those forms with large cells, which favored lower metabolic rates. The existence of low metabolic rates in some of the Late Paleozoic lungfishes is indicated by the evolution in Permian lepidosirenids of the ability to estivate. The modern lepidosirenid *Lepidosiren*, which has 121 pg DNA per haploid genome, and *Protopterus*, with 142 pg DNA per haploid genome, have a much

Most of the paedomorphic changes in lungfish evolution occurred in the Devonian Period.

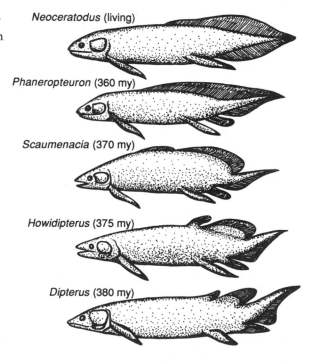

Neoceratodus (living)

Phaneropteuron (360 my)

Scaumenacia (370 my)

Howidipterus (375 my)

Dipterus (380 my)

larger cell size than the other living lungfish, *Neoceratodus*, which does not estivate (*Neoceratodus* has just 80 pg DNA per haploid genome). Moreover, an analysis of osteocyte volume in *Neoceratodus* has shown no increase in cell size over the last 250 million years. So, like pletho-dontid salamanders, the evolution of larger cell size in lungfish was associated with a wide range of paedomorphic changes.

We are only just beginning to scratch the surface in our under-standing of the interplay between factors such as paedomorphosis, cell volume, genome size, and metabolic rates. Only with further studies in other groups of organisms will we be able to start to unravel the intricate interrelationships between these various factors and come to grips with their role in evolution. What about evolutionary changes in cell size in other groups of organisms? One group that should obviously be investigated is the Foraminifera, single-celled protists that have an extensive fossil record. It is well known that many lineages of Tertiary foraminifers show trends of size increase. Since these organisms are single celled, this means that such increases in size of the whole organism also show an increase in cell size. How-ever, little is known concerning the correlation between such cellular changes and developmental rate in foraminifers. If cell and genome size in these protists was also indirectly correlated with developmen-tal rate, as in the amphibians and lungfishes, then we would expect to find numerous examples of paedomorphosis in the fossil record. What little work has been done suggests the possibility that this style of heterochrony may be common.

Cell Death and the Evolution of Cat Brains

Living deep within the Sierra Morena in the southern part of central Spain is a cat that appears to be an evo-lutionary "hanger on": a remnant of the Late Pleisto-cene fauna that once roamed southern Europe. Su-perficially similar to a domestic tabby cat, although appreciably larger, this cat is characterized by the na-ture of its teeth, having particularly large canines and premolars. It is also the possessor of a much larger brain than the domestic cat. Robert Williams of the University of Tennessee at Memphis, along with Carmen Cavada and Fernando Reinoso-Suárez of the Univer-sidad Autónoma de Madrid, analyzed the comparative evolution of the visual system of this Spanish wildcat (*Felis silvestris tartessia*) and the domestic cat (*Felis catus*) in order to establish the range, rate, direction, and magnitude of evolutionary change between the two species.[11]

Morphologically, the living specimens of the wildcat so closely resemble Pleistocene (about 15,000-year-old) specimens that they consider this form to have avoided the closing Pleistocene catastrophe, which wiped out so many larger mammal species around the world (apart from Africa). Indeed, they are of the opinion that this particular wildcat has more in common with the extinct Pleistocene wildcats than with other living wildcats in the species, *Felis silvestris*. The domestic cat is thought to have evolved from *F. silvestris* stock in the Late Pleistocene by a reduction in body size via the wild subspecies *F. silvestris silvestris* and *F. silvestris lybica*, some 3,000 years ago. The decline in body size predated domestication and parallels similar size reduction in other mammalian species at the close of the Pleistocene about 20,000–12,000 years ago. This size reduction is thought to be associated with the postglacial climatic change and human competition. The second phase of size reduction was associated with domestication, resulting in changes in brain size and behavior. Of particular significance in the study by Williams and his colleagues was the determination of the factors that may have been responsible for the reduction in body size and brain size in the domestic cat. Was it a case, like the amphibians and lungfishes, of a simple relative reduction in cell size, or were other factors involved? The smaller body and brain size of the domestic cat indicate that paedomorphic processes were at play. But what could have been driving such changes at the cellular level?

In order to determine the cellular factors responsible for distinctions in body size and brain size between the two species, Williams and his colleagues analyzed populations of cells in the retina and other parts of the eye. After all, the eye is part of the brain, so any differences in cellular construction in the eye should reflect overall differences within the brain as a whole. The brain of the Spanish wildcat reaches up to 37 g in the largest male. In domestic cats the brain is much smaller, averaging 27.6 g. There is an allometric relationship between brain size and body size. With an average twofold size difference between the two cat species (6–7 kg for the wildcat, 2.5–

3.5 kg for the domestic cat), there is a 25–30 percent difference between the brain weights, the brain of the wildcat being bigger. However, the rare domestic cats of huge proportions that turn up do not have brain sizes to match their bodies—they are all brawn and little brain. For instance, an enormous 9-kg domestic cat that Williams and his colleagues investigated still had a brain that weighed only just over 28 g.

The total ganglion cell densities are much higher in the retinas of the wildcat than in the domestic cat, varying from 50 to almost 100 percent more. This provides evidence to suggest that rather than cell size affecting the size of the brain, it is the number of cells present. Thus, the reduction in brain size during the evolution of the domestic cat occurred as a result of reduction in the number of cells. Like the 25–30 percent reduction in brain weight, there is a similar reduction in the number of cells. In other animals, such as some rodents, rapid evolutionary size reduction is known to have been associated with reduction in cell number. However, long-term brain evolution over millions of years appears, in this group at least, to have involved changes in both cell size and cell type.

But perhaps the most significant finding that came out of the work of Williams and his colleagues was not just that the differences were due to cell number, but the mechanism for that difference. Interestingly, both species of cats had the same gestation period, 63 days, and birthweight was the same (around 100 g). They suggest that rather than variations in the rate of cell growth causing the distinctions in brain size, the same number of cells may actually be generated during embryonic development, but during subsequent development far more die in the domestic cat than in the wildcat. Williams and his colleagues have found that cell death in the retina of the domestic cat is particularly high during prenatal and postnatal development, with up to 80 percent of all ganglion cells dying. This is much more than in, say, primates, including humans. Perhaps, then, the large number produced initially occurs as a result of an evolutionary heirloom from the larger ancestor. This fits in with the finding that changes in cell number operate over short-term evolutionary change, whereas changes in cell size may have more impact over much larger tracts of geological time, involving tens to hundreds of millions of years.

Escape to
Miniaturi-
zation

In 1989, I paid a visit to a small limestone quarry in southwest Australia, near the town of Albany. I hadn't visited the quarry in more than a decade, and my aim was to collect fossil sea urchins. I had never had much luck at this place on my previous visits. As it turned out, I had been looking in the wrong place. This time, a fortuitous gust of wind set me on the track of finding one of the smallest sea urchins that has ever lived. Accompanying me on this trip was my four-year-old daughter Katie, equipped with her tiny hammer and plastic bag in which to put her fossils. As we set off to the main workings, a gust of wind suddenly blew Katie's bag out of her hand and carried it off in the direction opposite to where we had intended to go. Bending down to pick up the bag where it had landed on a pile of old, deeply weathered limestone, left for years by the quarrymen, I saw an urchin. Then I saw another, and then another. We had hit a treasure trove of urchins, perfectly weathered out of their limestone tombs.

Most of these fossils were a variety of cassiduloid urchin known as *Echinolampas.* These particular urchins are not like the sort that most people think of when you mention urchins—those prickly balls whose main object in life seems to be to embed their spines as deeply as possible into your feet as you unsuspectingly walk over a rock platform. No, these urchins are not the so-called regular variety, but the group containing heart urchins and sand dollars, the so-called irregular urchins, furnished with a dense coat of fine, hairlike spines. Most of the specimens I had been collecting were about 20 to 30 mm long. As I bent down to pick up one particularly well-preserved specimen, I was amazed to see, sitting primly on the top, like Peter Pan on his pedestal, a tiny urchin, less than half the size of a grain of rice. My first reaction was to think that this was a juvenile urchin of the *Echinolampas.* This was not an unreasonable assumption, given its shape. Realizing that if there was one of these tiny urchins there had to be more, I adopted the typical paleontological bloodhound pose (nose to the ground, posterior in the air) and quickly started finding more of them. When I cleaned the specimens and looked at them in detail I was startled to find that these were not juvenile *Echinolampas,* for each specimen sported four genital pores—a sure indication of their adult status. What I was dealing with were minute adults of quite another order of urchins, the Clypeasteroida, or sand dollars. This realization made finding that first one sitting on a cassiduloid urchin even the more remarkable, for it has long been argued that the first sand dollars, which evolved as (relatively) recently as about 55 million years ago, may well have evolved from cassiduloid echinoids by progenesis. The similarity between these earliest sand dollars and juvenile cassiduloids is quite striking, hence my initial interpretation. Reaching no more than about 4 mm in length, this tiny sand dollar is even more remarkable for its method of reproduction. On its undersurface, just in front of the mouth, are two depressions. Inside these its own young would have been brooded. Most urchins, like many marine invertebrates, reproduce externally, with eggs being fertilized outside the body and the juveniles becoming part of the free-swimming plankton. But a few produce larger eggs that remain on the surface of the urchin's body, either nestled between the spines or in special brood pouches, where they are fertilized. Here the young urchins develop, well protected, until they are large enough to leave their mother. Today, such direct development is relatively common in marine invertebrates, including many different types of echinoderms that inhabit the waters around Antarctica. I will be discussing the significance of this more in Chapter 10.

Forms like the tiny *Fossulaster* retain a number of characters present in ancestral juveniles: small size, fewer spines, and fewer tube feet, and illustrate the importance of progenesis as a mechanism for inducing the evolution of a major new group of animals. Not only does the early progenesis and evolution of a very simple body plan produce a pronounced morphological difference between the adult descendant and the ancestor, but there is likely to be a major adaptive shift into a substantially different niche. While the genetic difference will initially be minimal, the pronounced niche separation will ensure genetic isolation and divergence.

This method of achieving paedomorphosis involving the evolution of very small body size and simple morphology has been proposed as a significant process in the evolution of other major new groups of animals. For instance, Simon Conway Morris of the University of Cambridge and John Peel of the University of Uppsala in Sweden have recently suggested that peculiar, armored, sluglike halkieriids, which have small shells covering their head and tail, and which lived over half a billion years ago, may have given rise to brachiopods. This, they argue, could have occurred by paedomorphosis, early progenetic maturation essentially freezing the animal with its two shells, with the loss of the intervening sluglike body.[12]

Perhaps one of the most significant roles of paedomorphosis has been the evolution of insects. This is one of the groups that Gavin de Beer identified as having arisen by neoteny. Since Ernst von Baer in the 1820s noted the similarity between the early embryonic stages of some millipedes and centipedes and adult insects, the favored explanation for the evolution of insects has been that it occurred by a loss of body segments, combined with a loss of appendages from some ancestral centipede-like ancestor. For example, the millipede *Glomeris*, upon hatching, possesses a thorax of three pairs of appendages, followed by an abdomen of eleven segments with very reduced legs. As it grows and passes through a number of molts, the millipede increases its numbers of segments and limbs. It is argued that the evolution of insects from such a millipede or centipede ancestor would have involved the operation of paedomorphic processes. This was

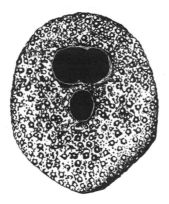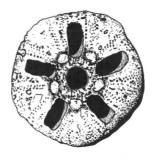

Some sea urchins brood their young directly, missing the planktic, free-swimming larval stage. Some even develop special pouches in which they brood their young. These can either be on the ventral surface in front of the centrally positioned mouth, as in the tiny (less than 10 mm long), 20- to 40-million-year-old *Fossulaster* (*left*) or as five pouches on the dorsal surface, as in the equally small 20- to 30-million-year-old *Pentechinus*.

Insects could have evolved from myriapods through intermediate forms like the extinct euthycarcinoids by a paedomorphic reduction in segment number, and then loss of appendages. Suppression of the *Hox* genes that regulate both segment and appendage formation could have produced such a change during the Early Paleozoic.

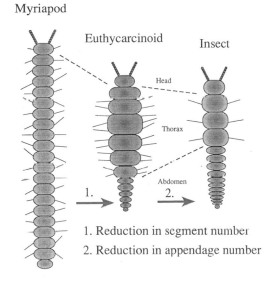

1. Reduction in segment number
2. Reduction in appendage number

likely to have been progenesis. The result is that adult insects resemble, in some respects, juvenile centipedes and millipedes.

However, a serendipitous find in 1990 by Kris Brimmell, of the Western Australian Museum, in Kalbarri National Park in Western Australia led me to suggest with my colleague Nigel Trewin from Aberdeen University that an alternative scenario could be envisaged. What we have claimed is that insects might actually have evolved from another group of arthropods, now extinct, called euthycarcinoids. Resembling, in many respects, a cockroach that had sprouted many more legs, euthycarcinoids were thought to have first evolved about 300 million years ago. That was until our Kalbarri find. This beast, since christened *Kalbarria brimmellae*, was 120 million years older, coming from 420-million-year-old Silurian rocks.[13] Significantly, this placed euthycarcinoids earlier in the fossil record than the earliest insect remains, which are found in Early Devonian sediments.

What euthycarcinoids represent, I think, are the "missing link" between centipedes/millipedes and insects. Like the centipedes and millipedes, they had more pairs of legs than insects (eleven pairs to be exact), but they resembled insects more than centipedes/millipedes in having fewer body segments (actually about the same total number as insects). Moreover, they had a body already subdivided into a thorax and an abdomen. Where the euthycarcinoids differed from the insects, apart from their greater number of legs, was in the position of the boundary between the head and the thorax. With two fewer segments in the euthycarcinoid head, what needed to occur for the insect head condition to be attained was the incorporation of two body segments into the head. Such a change is developmentally not very difficult. For instance, by analogy with modern arthropods, changing the concentration of a particular protein along the head/tail gradient in a euthycarcinoid egg would result in moving the position of the junction between the head and thorax. Coincident with this was a paedomorphic reduction in the number of appendages, and the resultant evolution of insects. Although we do not know the details of euthycarcinoid ontogeny, it is likely that the number of segments and the numbers of appendages increased through successive molts. Progenetic early maturation may have caused this reduction in number of appendages due to the failure of one set of genes to trigger homeotic genes to activate appendage formation. The evolution of such a major group of animals as the insects clearly shows how just slight changes in the ancestral developmental program can have profound evolutionary implications.

However not all small, progenetic animals gave rise to whole new suites of animals. In many cases, taking the progenetic road to paedomorphosis, with its ensuing small body size and reduced, less complex structures, was a road to specialization for a particular niche. Examples of extremely small body size are widespread in most, if not all, groups of animals today, as well as in extinct groups for the last half a billion years. In living animals examples have been described from a range of worms, brachiopods, mollusks, arachnids, crustaceans, insects, echinoderms, fishes, amphibians, reptiles, birds, and mammals. Examples have also been described from extinct groups like trilobites, ammonites, and dinosaurs. The added advantage of studying fossil examples is that we can see sequences of increasingly progenetic forms making up lineages, leading to smaller and more simpler structures. Even among the earliest trilobites, from the Early Cambrian of Scotland, as I described in the Prologue, we can see that such a paedomorphic process was operating and functioning as a major evolutionary process half a billion years ago.

There is little doubt that body size places particular constraints on how organisms function, as well as their ecological place in communities. The evolution of extreme size reduction will affect not only the overall anatomy but also the physiology, life history, and behavior. But how small is small? Some of the more extreme small vertebrate paedomorphs can have body lengths as short as 10 mm, and weigh just a few grams. Invertebrates can be even smaller. One Antarctic species of bivalve has a maximum length of just 1.1 mm, making my fossil *Fossulaster* seem of almost gigantic proportions by comparison. The Antarctic realm, in fact, is a region of great contrasts when it comes to body size, with more than its fair share of extreme body size, both gigantic and minute species being relatively common. Indeed, it has been documented that 61 percent of bivalve mollusks are sexually mature at lengths of less than 10 mm. Elsewhere, entire communities of minute organisms exist, the so-called interstitial or meiofauna.

The environment is all a question of scale. While to us the wet sand upon which the gentle waves lap may be a useful medium for building sandcastles, to certain organisms each of these grains of sand takes on monstrous proportions, as a great boulder to negotiate around; to hide under; or to avoid being crushed by. Inhabiting the spaces between sand grains in most aquatic environments, especially the marine, is another world—a lilliputian community of tiny organisms. Most invertebrate groups are represented, and there are whole groups of animals that occur almost exclusively in this interstitial environment. It is within this fauna that the world's tiniest animals live. For example, there are hydroids just 1 mm in length and some worms barely a third of a millimeter long. Other members of this fauna are giants 3 mm long. A half to one millimeter would seem to be the lower limit that most animals can attain and still function effectively.

The paedomorphic nature of this meiofauna is shown not only by the very small body size but also by the reduced nature of many of their morphological characteristics. For instance, the hydroid *Halammohydra* has far fewer tentacles than other larger hydroids that do not inhabit this microscopic world. It has just seven tentacles, whereas ancestral hydroids have up to 36. In the 1950s Bertil Swedmark of the Kristeneberg Zoological Station in Sweden demonstrated the role of paedomorphosis ("neoteny," as he called it) in one such member of the meiofauna. He showed how in what he termed the "aberrant" (merely different) polychaete worm family Psammodrilidae, one species, *Psammodriloides fauveli*, is 25 times smaller than another called *Psammodrilus balanoglossoides*. Not only that, but its adult anatomy resembles juveniles of *Psammodrilus*. He further showed that the number of cells in this paedomorphic form is about 25 times less than in the nonpaedomorphic *Psammodrilus*. Other interstitial organisms, including worms, cnidarians (corals), harpacticoid copepod crustaceans, and hydroids, all retain ancestral larval features as adults.[14]

At less than half a millimeter long, one of the smallest known animals, the archiannelid *Nerillidium troglochaetoides*, towing an embryo.

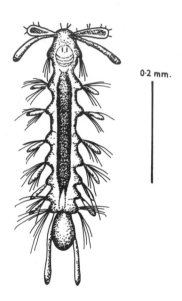

0·2 mm.

In terms of heterochrony, such small paedomorphs can evolve in two ways: by progenesis or neoteny. In the former process the incidence of premature maturation will almost invariably be associated with smaller size. In the case of neoteny it is less simple. Neoteny as a process, in other words a reduced rate of development, usually affects only certain structures. For neoteny to produce a small paedomorph, all growing parts of the entire organism would have to reduce their rate of development, and reduce the amount of size increase. Few studies of living animals have been carried out to ascertain the relative frequency of these two mechanisms, let alone for fossil groups.

The impact of paedomorphic size reduction usually results in a reduction in structural complexity, as in the meiofauna. Often only certain structures of an organism will show paedomorphic reduction in size and complexity. This will occur by either the reduced degree of development of particular structures (e.g., the limbs in "legless" lizards) or the complete loss of structures, such as appendages in snakes. For example, in one particular species of salamander, called *Thorius*, the adult skulls have fewer bones than in larger, nonpaedomorphic genera. Of the remaining bones, some are so reduced that they do not connect with other bones. The result is that the brain is, to a large degree, unprotected by the bones. While these paedomorphic forms, as adults, resemble juveniles of their presumed ancestors, they are not exact matches. While most structures will be reduced, a few may be proportionately relatively large. This is because structures adapted for a tiny juvenile may not be appropriate for a slightly larger adult. Progenetic adults, although very small, are likely still to grow larger than corresponding ancestral juvenile forms, to some degree. One of the reasons for the success of these miniature Peter Pans, other than the ability to occupy, as an adult, a niche that is quite different from that occupied by the ancestral adult, is that often morphological novelties arise as a consequence of the progenesis. For example, tiny phallostheid fishes from Southeast Asia have evolved a peculiar copulatory organ developed from the pelvic fins. In another case, minute adult featherwing beetles have special ciliated wings, only a quarter of a millimeter long.

The occurrence in both living and fossil faunas of whole suites of miniature, paedomorphic forms suggests that the environment can have a direct influence on inducing the selection of such forms. In addition to the minute meiofauna living between sand grains, the fossil record shows that under certain environmental conditions in the past, particular faunas have been dominated by small paedomorphs. Many marine invertebrate so-called dwarf faunas have been documented from many parts of the fossil record, from Ordovician rocks 450 million years old through to the present day. These faunas may consist of entire suites of brachiopods or mollusks and are often found in sedimentary environments that are characterized by soft, unstable substrates. Ernest Mancini of the University of Alabama has interpreted the small body size and simple morphologies of 70-million-year-old Late Cretaceous oysters and ammonites found in Texas as having arisen from paedomorphosis. This he attributes to progenesis. He considers that the advantages gained by the resultant small size are that, unlike larger shells, these small shells could occupy soft, unstable sediments.[15] Likewise, the many very small, progenetic brachiopods occurring in Late Cretaceous chalks in Europe were similarly living on very soft sediments that were unable to sustain the weight of larger shells. A number of these brachiopods had lost any ancestral attachment mechanism, such as a pedicle. However, in some, a pedicle is present in juveniles, but is closed in adults, which rested on the soft sediment on a convex valve. Others, among the forty-three species of tiny brachiopods described from the one fauna, would have lived attached to small sponges and bryozoans or any shell debris lying on the soft sediment.[16]

Dieter Korn of Tübingen University in Germany has shown how some 370-million-year-old Late Devonian ammonoid faunas are dominated by small paedomorphs (such as *Balvia, Linguaclymenia, Parawocklumeria*). They occur in association with juveniles of nonpaedomorphic ammonoids. Other faunas are dominated by larger, nonpaedomorphic forms. Differences in the occurrence of the ammonoid sizes can be correlated with different ecological conditions. Several independent evolutionary lineages are characterized by a rapid size reduction. These dwarfs generally resemble the ancestral juveniles morphologically but are not scaled-down versions of their normal-sized ancestors. Korn considers that the size reduction probably resulted from a drastic shortening of ontogeny by progenesis. Significantly, these tiny ammonoids with very simple shell morphologies formed the basis for adaptive radiations that led to new types with various shell and ornament peculiarities, such as triangular coiling, apertural projections, ventrolateral grooves, and different sculptures with ribs.[17]

Korn has suggested that the occurrence of tiny progenetic shells arose from unstable environmental conditions. It is possible that the paedomorphosis in the ammonoids may have been induced by temperature changes, the late Devonian paedomorphs occurring in a shallower, and probably warmer, environment than the nonpaedomorphic ammonoids. Alternatively, the progenetic event might have been a product of environmental instability arising from falling sea levels, requiring a faster reproduction rate, achieved by precocious maturation. The drastic size reduction caused by progenesis was

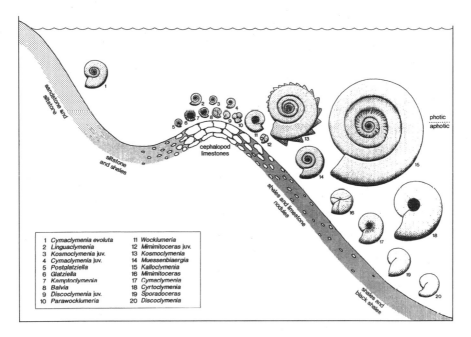

1	Cymaclymenia evoluta	11	Wocklumeria
2	Linguaclymenia	12	Mimimitoceras juv.
3	Kosmoclymenia juv.	13	Kosmoclymenia
4	Cymaclymenia juv.	14	Muessenbiaergia
5	Postglatziella	15	Kalloclymenia
6	Glatziella	16	Mimimitoceras
7	Kamptoclymenia	17	Cymaclymenia
8	Balvia	18	Cyrtoclymenia
9	Discoclymenia juv.	19	Sporadoceras
10	Parawocklumeria	20	Discoclymenia

Panorama of Late Devonian ammonoid habitats. Most of the small ammonoids (both paedomorphs and juveniles of nonpaedomorphs) lived mainly in elevated areas in shallower water, whereas the larger adults of nonpaedomorphs occurred in deeper water.

achieved at a geologically very rapid rate. As Korn has noted, the response to selective pressure on smaller body size was probably achieved with only minor genetic change. The outcome was the evolution of tiny ammonoids which differ remarkably from the morphology of their adult ancestors.

Environmental factors can also select for paedomorphosis under situations where only certain features, such as eyes or limbs, are affected. With the dwarfed faunas, the entire organism was affected. However, in cave environments, many of the organisms that show adaptations for living in such an environment are able to do so because only some anatomical features are targeted. The often bizarre structures of cave-dwelling animals have been the subject of much debate. Often considered to be the product of what has been termed "regressive" evolution, these organisms are anything but. Although specific structures, like eyes, might be reduced or lost, such evolution is not a backward step. It is a specialization that enables a very specific environment to be inhabited. If we are to cast epithets describing degrees of evolution, then these subterranean animals would be considered to be "highly evolved."

Recent research is indicating that living in a stressful environment, such as a deep cave, might actually induce morphological changes. One way that organisms cope with high levels of environmental stress is to decrease their metabolic rate, and many cave-dwelling animals are known to have lower than normal metabolic rates. Shifts to new habitats may in fact be facilitated by stress-induced paedomorphosis. This stress directly affects hormone production, slowing development and producing paedomorphic characters. For instance, in amphipods, exposure to light can have a direct effect on antennal gland hormones that are important in regulating development. The most paedomorphic of these amphipods is a cave-dwelling species, *Spelaeorchestia koloana* from Kauai. Related species that inhabit dark, leaf litter habitats are less paedomorphic, but more so than their ancestors that inhabited more strongly lit environments.[18]

Among cave-dwelling vertebrates, paedomorphosis is best known in plethodontid amphibians. Of the nine species of cave-dwelling salamanders, all, bar two, are paedomorphic and retain larval gills throughout their life. The larval nature of these animals has been recognized for a long time. While the most celebrated cave salamander, *Proteus anguinus*, was the first cave species to be scientifically studied and described, by J. N. Laurenti in 1768, it was recognized long before that: there is a stone carving, dating from the tenth or eleventh century, of a *Proteus* on a wellhead. And *Proteus* that still lives today in the caves at Postojna in Slovenia was considered by Johann Weichard von Valvasor, writing in 1689, to prove the existence of dragons. According to him, floods from the caves were caused by a dragon who lived there. The *Proteus* washed out were thought to be the dragon's larvae.

Paedomorphosis has undoubtedly been a major trend in the evolution of many groups of organisms. Evolution is not a case of trends to ever-increasing anatomical complexity; simplification can be equally as potent in opening up new adaptive pathways. Perhaps the most extreme effect of progenesis in some groups of organisms has been the adoption of a parasitic mode of life. This is probably best exemplified in tiny worms called Acanthocephala. Measuring usually just a few millimeters in length, the attainment of a parasitic mode of life has been attributed to progenesis by Simon Conway Morris and David Crompton of the University of Cambridge.[19] Acanthocephala are found most often living in teleost fish, but they also dwell deep within the guts of birds, mammals, and the occasional reptile and amphibian. From a large free-living ancestor, perhaps resembling worms like *Ancalagon minor* from the 530-million-year-old Middle Cambrian Burgess Shale in British Columbia, tiny parasitic worms, lacking a gut and teeth, may have evolved via small meiofaunal species by progressive progenesis. Like the meiofauna, the evolution of parasitic acanthocephala was achieved by both miniaturization and anatomical simplification.

The advantage of such extreme simplification lies not just in the small size (a functional prerequisite for parasites), but also in high fecundity, with specialization of the reproductive organs to maximize

The blind cave salamander
Proteus anguinus from
Slovenia

egg production. Again here we see how progenesis can lead to an overall reduction in size and complexity in most anatomical features, but to an increase in another. As we shall see in the next chapter, such developmental trade-offs have, I believe, played an important role in many major steps in evolution.

8

Images of the Past, Shapes of the Future

Organic Life beneath the shoreless waves

Was born and nurs'd in Ocean's pearly caves;

First forms minute, unseen by spheric glass,

Move on the mud, or pierce the watery mass;

These, as successive generations bloom,

New powers acquire, and larger limbs assume;

Whence countless groups of vegetation spring,

And breathing realms of fin, and feet, and wing.

Erasmus Darwin, *The Temple of Nature*, 1803

IN THE EARLY SUMMER OF 1994 THERE OCCURRED A HUGE MASSING OF emus along the vermin-proof fence in Western Australia. Designed to keep dingos and rabbits out of the agricultural land, this fence trapped tens of thousands of scrawny, starving birds as they attempted to migrate out of the drought-stricken, semiarid region of the state into the relatively more lush agricultural land toward the coast. By the time they reached the fence the emus had already traveled many hundreds of kilometers—carried over dry mulga scrub and spinifex desert on long, powerful legs. These legs, and the huge body they supported, had served members of the species well, as they have members of other species of giant flightless birds that have roamed the southern continents for tens of millions of years. But the very legs that had evolved in these particular ratites (as giant flightless birds are known) did so at the expense of the other paired limbs—the wings. For at some stage, far back in their evolutionary history, the ratites underwent three major morphological changes: an increase in body size and weight; a decrease in wing and sternum size (to the point where they were incapable of flight); and a very great increase in leg size.

Ironically, as the emus slowly died, eagles and other raptors circled high overhead, gliding silently high over a barrier that to them did not exist, before swooping down to feed voraciously on the ever-increasing pile of carcasses. While their huge, powerful legs had served them well in the endless Australian bush, the loss of their wings turned out, at this time and at this place, to be the emus' undoing. As their lives slowly ebbed away in the searing heat, the last sight that passed before the eyes of many of these birds would have been of the descendants of their ancient, flighted ancestors sailing effortlessly and smugly over this fence of death.

A Little Bit of Give and Take

While there are many species of flightless birds living today, far and away the most spectacular are the ratites. Contained within this group are emus (Australia), ostriches (Africa), rheas (South America), cassowaries (Australia and Papua New Guinea), and kiwis (New Zealand), while in the relatively recent past, but now extinct, were the moas (New Zealand), dromornithids like *Genyornis* (Australia), and elephant bird, *Aepyornis* (Madagascar). With the exception of the rather diminutive kiwis, ratites are large, flightless land birds taking their name from the flat, raftlike shape of their sterna. Although some early naturalists were of the view that the widespread distribution of ratites on the southern continents suggested that they evolved from independent stocks, plate tectonic theory suggests that they had a common origin in the Mesozoic on the great southern continent of Gondwana.

Ratites combine some extremely primitive features (which suggest an early divergence from the ancestors of modern birds) with a potpourri of highly derived, often unique characters. Originally they were grouped together on the basis of the absence of a keel on the sternum, in combination with short wings. But ratites also possess certain features not found in any other birds, such as a "loose" plumage structure, fusion of the shoulder blade and coracoid (the bone that links the sternum and shoulder blade), substantially reduced wings, and powerfully developed pelvic limbs. Many early naturalists, Charles Darwin among them, attributed the greatly reduced wing, which was useless for flight, to "disuse" or "degeneration." Others interpreted the "degenerate" characters of ratites as being the products of truncated development.

With the decline of the concept of recapitulation in the twentieth century, and the rise to dominance of paedomorphosis as an explanatory concept, ratites became one of the *causes célèbres* for this view. Its champion was Gavin de Beer. In 1956, writing in his major work on the evolution of the ratites, de Beer was firmly of the opinion that "the presence in the Ratites of nestling-down, permanent sutures between the bones of the skull, and the dromaeognathous [paleognathous] structure of the palate [were] demonstrably the result of neoteny or the secondary retention of features which were juvenile in the ancestors of the Ratites."[1] And two years later, in *Embryos and Ancestors*, de Beer observed that "it is clear . . . that the ostrich, in retaining in the adult a type of plumage characteristic of the young of other birds, is neotenous."

So, it was not only the flightless nature of these birds that attracted the paedomorphologists; the idea was extended well beyond the wings to aspects of the cranium, the pelvic girdle, and the feathers. The numerous scientists who argued for the paedomorphic nature of ratites called it "neoteny," so beginning a tradition of equating neoteny with the retention of juvenile characters. This became entrenched in the biological literature, to such an extent that even today confusion is still created by some biologists who use this term in a descriptive sense. Now this manifestation of juvenile characters is called "paedomorphosis," the term "neoteny" being retained just for the particular process that leads to a reduced rate of development, and which *results* in paedomorphosis: neoteny the process, paedomorphosis the effect.

The mental contortions that some biologists went through in the 1950s in order to strait-jacket every feature into "neoteny" is shown by de Beer's explanation of the evolution of the palate in ratites. In his work on the evolution of ratites published in 1956, de Beer wrote:

> For those, if there be any, who still believe in the theory of recapitulation, it would no doubt be tempting to say that the neognathous palate "recapitulates" in its development the condition of the paleognathous palate which would therefore be ancestral. But in view of the overwhelming evidence that the Ratites are secondarily descended from flying birds . . . [and] the fact that the Ratites already show neoteny in two other features . . . it is impossible to believe that in their palates the Ratites are primitive.

So, in other words, because a handful of features are paedomorphic, the entire creature has to be seen as a product of paedomorphosis. And this view is still, by and large, entrenched in current evolutionary biological thinking. Even though few would now argue against both paedomorphosis and peramorphosis as being valid het-

The extinct flightless ratite, the moa from New Zealand. This species, *Dinornis maximus*, was the largest of a number of species of moas. At its feet is the living flightless kiwi.

erochronic effects, organisms still tend to be viewed as either paedo-morphic or peramorphic. Yet, as I shall argue repeatedly in this and subsequent chapters, most organisms are a cocktail of paedomorphic and peramorphic features. Indeed, it is this very potent mixture, in such a wide variety of proportions, that has contributed significantly to the vast panoply of organisms that have inhabited this planet for the last half a billion years.

There is little doubt that some aspects of ratites are the product of paedomorphosis. The extent of this has recently been reviewed in some detail by Brad Livezey of the Carnegie Museum in Pittsburgh. For example, the tiny wing bones and delayed closure of cranial sutures of ratites may well have evolved by neoteny, as it is known that closure of cranial sutures can take many years in these birds.[2] As I discuss later in this chapter, many instances of delays in the transition from one ontogenetic stage to another (in this case from unclosed to closed cranial sutures) can be viewed as resulting in paedomorphosis, as an earlier growth phase is extended in the descendant. However, if this phase is undergoing pronounced allometric growth in other characters, peramorphosis might ensue. Many other features in ratites clearly owe their origins to a relative peramorphic increase in the degree of growth. These include the large body size (arising from a delay in the onset of maturity and/or an acceleration in growth rate of the birds as a whole) and the relatively very large legs. What happened in the evolution of ratites is particularly instructive, because it parallels what has happened in many other vertebrates—an evolutionary trade-off. If selection has strongly favored a great increase in certain structures, then this will often have been at the expense of others. After all, what is most important to an ostrich—its pathetic, useless wings, which have so attracted the attention of evolutionary biologists, or its big body and massive legs, which can disembowel a hungry lion with a single kick? The latter, I think. The paedomorphic features are the evolutionary trade-offs: in the case of ratites, features such as loss of wings and quilled feathers. A growing organism has only a certain amount of energy to expend on growth. If selection favors individuals that have concentrated growth into certain structures, then unless there is an overall increase in cell size and cell number, some areas have to lose out in a development trade-off.

So, in ratites, we see what Brad Livezey has termed "a complex constellation of variably localized heterochronic changes in skull, pectoral [wing] and pelvic [leg] appendages." Depending on the availability of suitable ecological niches, the ecological requirements will dictate the ingredients of the cocktail. Thus, at the beginning of the Cenozoic, immediately following the demise of the dinosaurs 65 million years ago, there was a gaping, empty large-predator niche. This was not filled immediately by mammals, but by large, flightless birds, such as *Phororachus*. Possessing not only large bodies and powerful legs, they were also blessed with an enormous head and massive beak, the product, no doubt, of their hypermorphic or accelerated growth, dragging along ancestral allometries.

As Livezey has noted, most cases of flightlessness in birds are associated with an increase in body size, perhaps implying that simple acceleration in leg size alone does not provide sufficient selective advantage. To produce a *really* big leg, one that can allow a few hundred kilograms of ostrich to outrun a Linford Christie, there was a developmental imperative to increase the body size. No large body, no large legs. So by delaying the onset of maturity, leg size was proportionately increased as body size increased. For instance, R. McNeill Alexander of the University of Leeds in England has shown that the enormous, powerful legs possessed by the extinct moas occurred simply by such an extrapolation of ancestral growth allometries as body size increased.

A similar pattern of heterochronic give-and-take also occurs in flightless carinate birds. Traditionally called "carinates" because of the presence of a keel or carina on the sternum, many groups display the tendency to flightlessness. Examples occur in cranes, rails (which account for about one-third of all known flightless carinates), all penguins, some grebes, waterfowl, and several genera of auks, and there are even flightless cormorants and parrots. Thus, for instance, within the Galápagos cormorant, peramorphic enlargement of the pelvic limb and of body size is particularly noticeable, in association with a "compensatory" shift in specialization of the appendages. While the presence of wings in seabirds like cormorants is usually advantageous in catching food, moving between feeding areas, and locating nesting sites, in the case of the Galápagos cormorant, food resources are to be found all year round close to the islands on which they nest. The effect of El Niño conditions in inducing unpredictable variation in food resources has militated against predictable seasonal migration. Brad Livezey has argued that over 5 million years of evolution the result has been not just the loss of flight by paedomorphic wing reduction but also the peramorphic increase in leg and body size. As Livezey stresses, this trade-off in appendage development arose from what he calls "developmental economy," for functional wings would have been of little benefit in such an insular environment. Furthermore, larger body size would have been beneficial when foraging. These features have developed under unusual ecological conditions, where prey size is large, foraging ranges are restricted, survival of juveniles is high, and the developmental period prolonged.[3] We shall see in Chapter 12 exactly how the evolution of another vertebrate, one *Homo sapiens*, parallels the Galápagos cormorant with an almost frightening degree of similarity.

The Dodo— Giant Chick or Overdeveloped Adult Pigeon?

One of the earliest writings that recognized heterochrony in birds was made by H. Strickland and A. Melville in 1848 in relation to the extinct, flightless dodo (*Raphus cucullatus*). They described this icon of extinction as having

wings too short and feeble for flight, the plumage loose and decomposed, and the general aspect suggestive of gigantic immaturity. We cannot form a better idea of it than by imagining a young Duck or Gosling enlarged to the dimensions of a Swan. It affords one of those cases . . . where a species, or a part of the organs in a species, remains permanently in an underdeveloped or infantile state. Such a condition has reference to peculiarities in the mode of life of the animal, which render certain organs unnecessary, and they therefore are retained through life in an imperfect state, instead of attaining that fully developed condition which marks the mature age of the generality of animals. . . . The Dodo is (or rather was) a permanent nestling, clothed with down instead of feathers, and with the wings and tail so short and feeble, as to be utterly unsubservient to flight.[4]

So, even in an age when the prevailing view was one of recapitulation, the overt paedomorphic nature of the dodo was quite apparent, in the reduced wings and juvenile plumage. But, like the ratites, combined with this "infantile" appearance were peramorphic features: the huge body, relatively enlarged legs, and massive skull. Once again this combination is there in adults of this bird: some obvious juvenile traits in conjunction with enlargement of other features, including body size. Now recognized as a gigantic, flightless pigeon, the heterochronic trade-offs displayed by the dodo were mirrored in another extinct, flightless bird, the solitaire, which used to inhabit the Mascarene Islands in the Indian Ocean. Both the dodo and the solitaire exhibited profound sexual dimorphism, possibly the greatest ever known in carinate birds, arising, perhaps, from their extreme size. While we can argue endlessly about the adaptive significance of the various morphological features of the dodo that influenced its evolution and survival, until *Homo sapiens* ensured its immortality by bludgeoning the species to extinction, we must turn to aspects of its life history to help explain its particular cocktail of characters.

The extinct flightless giant pigeon, the dodo. Wings may be a paedomorphic character, but the large body size is a product of peramorphosis.

I shall discuss the relationship between heterochrony and life history strategies in more depth in Chapter 10. However, it is perhaps opportune to show briefly here how selection of particular suites of heterochronic features arose, not in response to the morphofunctional significance of these traits, but as a reflection of aspects of the life history. An organism's life history includes attributes such as life span, body size, number of offspring produced, and stability of the environment inhabited. One characteristic of animals that have evolved by hypermorphosis is that they typically have long life spans, large body size, produce few offspring, and live in stable environments. And so it is with dodos; for, as Brad Livezey has shown, dodos would appear to have been slow to mature, seemingly had a long life span, and produced just one egg. In all likelihood their predator-free environment, before humans came on the scene, was stable.

So perhaps selection of these hypermorphic features in this particular, strange pigeon, combined with the compensatory paedomorphic reduction in wings, occurred because selection was favoring the dodo's particular life history strategy. The relatively small, underdeveloped wings incapable of permitting flight were not, however, a maladaptive feature. While the evolution of flight in birds in the first place was probably a response to predation pressure, it mattered little to the dodo on Mauritius whether it could fly or not, because before human predators arrived on the island it was free from predation pressure. This one paedomorphic feature certainly would not have been a principal target of selection. Like the ratites, the dodo was a unique cocktail of both peramorphic and paedomorphic characteristics. However, the importance of the peramorphic attributes far outweighed the significance of the useless, paedomorphic wings.

The Evolution of Avian Flight

While ratites may have overcome predation pressure by the peramorphic evolution of large body size and legs, the evolution of flight in birds in the first place evolved, paradoxically, by peramorphic evolution of the wings. But, as in the ratites, this occurred in combination with the paedomorphic reduction of features other than the wings. The evolution of such major evolutionary novelties in many groups of animals that allowed the exploitation of new niches, arose, I would suggest, by heterochronic trade-offs—by mixing the paedo/peramorphic cocktail in the right proportions, at the right place, and at the right time.

There is a general agreement that birds evolved from theropod dinosaurs, although there is some debate as to which particular theropod groups are most closely related to birds. The most favored candidates are deinonychosaurian coelurosaurs. Tony Thulborn of the University of Queensland has proposed that birds can be thought of, in some respects, as paedomorphic theropod dinosaurs.[5] He based this supposition not only on the small size of the first birds (the Jurassic *Archaeopteryx* from the Jurassic of Germany and *Confuciusornis* from rocks of similar age in China) but also on the idea that feathers may have been present on juvenile theropods, being used as an insulating blanket. Recently, a feathered dinosaur has been found in China.

There are more concrete ways in which *Archaeopteryx* resembles a juvenile theropod. The most obvious is the shape of the skull and orbits. In young theropods such as *Coelophysis*, the eye was relatively very large, decreasing in relative size during ontogeny. Adult *Archaeopteryx* retained a relatively large eye. Similarly, the brain case is relatively inflated in juvenile theropods, as it is in *Archaeopteryx*. Tooth reduction, and then eventual loss in later birds, is another paedomorphic character—likewise, the shape of the teeth. A recent discovery in the Gobi Desert in Mongolia by Mark Norell and colleagues from the American Museum of Natural History in New York, and the Mongolian Academy of Sciences and Museum of Natural History, of a nest of dinosaur eggs that also contained two embryonic dromaeosaurid skulls, thought to belong to *Velociraptor*, indicates that the shape of the teeth in early birds was also paedomorphic. In these embryonic dinosaurs the teeth are simple, peglike structures, unlike the more complex adult teeth, but very similar to the teeth present in early, primitive adult birds.[6]

Tony Thulborn has also argued that the relatively long hand bones, forelimbs, and foot bones in *Archaeopteryx* are also juvenile theropod traits. Yet, compared with a juvenile theropod the hand and arm bones are much larger, suggesting that a localized peramorphic enlargement has occurred. This would have been of special significance to the evolution of flight. Without it, it is hard to imagine how a wing, and flight, could have evolved. The parallel evolution of the wing in bats and pterosaurs also occurred by just such a localized acceleration in growth of finger bones (see Chap. 9).

Larry Martin of the University of Kansas has raised the question of the importance of heterochrony in bird evolution, especially in relation to determining the timing of sexual maturity in birds.[7] Modern birds have terminal growth, and most growth has been completed when the wings are functional and tarsal bones have fused with the tibia and metatarsals. In other words, the chicks undergo rapid growth in the nest, but upon fledging, growth essentially ceases. Larger adult size can only be attained by an acceleration of growth, relative to an ancestor, or by a prolongation of this rapid juvenile growth phase, or by a combination of the two processes. *Archaeopteryx* shows less skeletal fusion than in living adult birds, indicating that a peramorphic increase in skeletal fusion has occurred during avian evolution. As Martin has shown, a list of the unfused or poorly fused elements in *Archaeopteryx* is little more than a catalogue of the juvenile condition found in living birds. What this shows is that following evolution of birds in the Jurassic Period, the overall trend in bird evolution has been one of peramorphosis. Thus, a number of characters, such as pubic reflexion, ilium prolongation, and increased fusion, appear in the fossil record in much the same sequence as in ontogeny. Adult Mesozoic birds show characters occurring early in the ontogeny of living birds, such as retention of distinct sutures in the skull. Bones such as the sternum, interclavicle, and uncinate processes of the ribs in *Archaeopteryx* are not ossified. This same condition occurs in the juveniles of living birds. Distinctions occur between *Archaeopteryx* and Cretaceous birds, such as *Apatornis*, *Ichthyornis*, and *Hesperornis* in the ilium, ischium, and pubis. These are separated by well-defined sutures in *Archaeopteryx* but are fused in adults of the Cretaceous birds. This pattern reflects the overall trend of peramorphosis within bird evolution as a whole, in contrast to the original pera/paedomorphic cocktail that resulted in the evolution of birds from theropods.

First Foot Forward

If asked to nominate the most important step in evolution over the last half a billion years after the Cambrian Explosion, which saw the appearance of the major phyla, I would probably push for the colonization of land by animals and plants. Although the first possible evidence for life on land comes in the form of mats of fossilized bacteria and algae found in 1,200-million-year-old cherts in Arizona, the earliest evidence for plants gaining a tenuous foothold is revealed by spores found in 470-million-year-old Ordovician rocks in Libya by Jane Gray of the University of Oregon. These are very reminiscent of spores from living nonvascular, or so-called lower, plants, such as bryophytes (mosses and liverworts) and pteridophytes (ferns). To date the earliest evidence for terrestrial animals has been found by Greg Retallack and Carolyn Freakes of the University of Oregon in Late Ordovician fossil soils, about 450 million years old, from Pennsylvania. This evidence is indirect, coming, as it does, in the form of deep, vertical burrows ranging from 2 to 21 millimeters in diameter, that are thought to have possibly been made by soil-inhabiting animals, maybe millipedes. The first direct evidence for both vascular (or "higher") plants and animals has been found in recent years in a range of localities in England and New York State in rocks of latest Silurian age, about 400 million years old. Careful extraction of fossil residues from these rocks has yielded not only the remains of simple plants, like *Cooksonia*, but also fragments of a wide range of arthropods: tiny spiders and spiderlike trigonotarbids, mites, and centipedes.[8] Even older rocks, about 420 million years old, from the Murchison Gorge in Western Australia, have yielded a diverse fossil assemblage of trackways made by a range of larger arthropods: eurypterids, euthycarcinoids, scorpions, and centipedelike animals, all of which were walking out of water on sand flats.

Close on the heels of these largely amphibious animals came the tetrapods—the first land-dwelling vertebrates, in the form of primitive amphibians. The earliest known amphibian body fossils were found in east Greenland: *Acanthostega*, which had eight toes on its front and hind limbs,[9] and its seven-toed close relative, *Ichthyostega*. These fossils tell us that the evolutionary transition from fishes to amphibians was both smooth and orderly. Three fundamental changes occurred in this transition: the loss of gills and acquisition of air-breathing lungs; fundamental changes in the shape and construction of the skull; and changes in the limbs, necessary to support the body weight out of water. *Acanthostega* and *Ichthyostega* show us that limbs capable of supporting the body out of water evolved before air-breathing lungs, because these early amphibians still possessed gills. The changes in the skull and limbs provide one of the classic, but much unheralded, examples of the role of dissociated heterochrony: a cocktail of peramorphic and paedomorphic features combining in a potion so strong that terrestrial ecosystems never looked back. But more than just describing the importance of such a combination of the two heterochronic processes for macroevolution, the evolution of the tetrapods illustrates how critical this cocktail of changes in the animal's overall anatomy was when operating on single structures.

The earliest fishes, which lived about 470 million years ago, got by without paired fins, but some of these early jawless fishes (agnathans) soon acquired simple fin folds in the sides of their bodies. These were the pectoral and pelvic fins—the precursors of the arms and legs of higher vertebrates. The first fishes to evolve an internal shoulder girdle to support the pectoral fin were the armored osteostracans. The pectoral fin in these fishes seems to have lacked ossified fin bones, indicating that they were simple, cartilaginous fin rays. Muscle control would have been weak, allowing just simple up-and-down movements. The evolution of the pelvic shoulder girdle followed the evolution of jaws.[10]

The first jawed fishes (the gnathostomes) were the placoderms and acanthodians. In both, pectoral and pelvic girdles were present. How the pectoral fin evolved into the structure that would become the standard vertebrate arm and leg is well illustrated in the evolutionary sequence from chondrichthyan to simple ray-finned fishes to lobe-finned crossopterygian fishes, then finally to tetrapods. The evolutionary mechanism that facilitated this evolutionary trend was dissociated heterochrony, with peramorphic increase in some bones combined with a paedomorphic loss of others. As with ratite birds, developmental trade-offs of particular elements at the expense of others resulted in profound anatomical, and consequently functional, changes, allowing a radical niche change to occur, from swimming to walking—from water to land.

The primitive pectoral fin in chondrichthyans was supported by many bony rays. These comprised three regions, a leading area known as the propterygium; the middle mesopterygium; and a branching metapterygium. The heterochronic events that led to the evolution of the crossopterygian lobe fin were a paedomorphic loss of the propterygium and mesopterygium, combined with a compensatory peramorphic elaboration of the persisting metapterygium. In more advanced crossopterygians, like the osteolepiform fishes, the first metapterygial bone is called the humerus, the same bone that came to support the upper arm in all later tetrapods. Articulating with the squat humerus were the radius and ulna, as in tetrapods; but both bones were relatively small.

One of the great debates in vertebrate evolutionary biology is how the unjointed fin rays that extended from the ulna were replaced by the complex of bones that form the hand and foot bones and the digits in tetrapods. Although this transformation is striking, the differences between the humerus, ulna, and radius in a crossopterygian like *Eusthenopterus* and the same bones in the early amphibian *Acanthostega* are very slight. A principal factor in understanding how the elements of the tetrapod foot or hand evolved has been the elucidation of what is known as the metapterygial axis. This is the principal axis running through the foot and hand of cartilaginous condensations, along which the bones of the foot and hand develop. Classically, it had been thought that the fish fin radials and the tetrapod digits were homologous elements, developing from either side of a straight axis that run down the middle of the hand and foot. However, Neil Shubin and Pere Alberch, when at Harvard University, showed that the metapterygial axis in tetrapods was strongly bent anteriorly through the distal carpel bones, the result being that all the digits branch posteriorly from this axis.[11] Hold your hand up in front of you. Your metapterygial axis runs down the middle of your arm, through the center of the wrist and hand, before curving strongly away to the side, below the base of your thumb. Even so, Shubin and Alberch have argued that the digits that you use to turn this page are homologous with the posterior radials of a primitive fish's fin. In such a scenario we could invoke the role of dissociated heterochrony, arguing that there had been a paedomorphic reduction in the *number* of bony elements, but with a compensatory peramorphic expansion of those few that remained, to evolve into digits. Ah, if only things were that simple!

Resolution of this problem may have been provided in a recent article published in the scientific journal *Nature* by Paolo Sordino, Frank van der Hoeven, and Denis Duboule of the University of Geneva in Switzerland.[12] What Sordino and his colleagues have shown is that the digits are *not* homologous to fin rays but are distinct evolutionary novelties—so our nice, simple heterochronic explanation flies out the window. However, such a model of paedomorphic reduction in the number of elements of a structure, associated with a corresponding peramorphic increase in the remaining elements, is another version of the developmental trade-off that I described in ratite evolution. But in this case, rather than affecting different structures within the organism, it affects a specific structure. I shall illustrate below how such a process is important in macroevolution. Sordino and his colleagues' radical model for digit development illustrates an even more intriguing heterochronic scenario, one that crops up in a wide diversity of organisms and has played an important part in the evolution of many organisms, from trilobites to humans. In the case of the evolution of tetrapods it involves changes in the timing of induction of different types of bone.

Unlike gastropods, with their external shell, or ants or spiders, with their hard outer skeletons, we are pretty unfamiliar with our skeleton, being, as it is, nicely tucked up in layers of muscle and skin. This is our endoskeleton. But we are much more familiar with the few remaining elements of our exoskeleton—an outer, bony covering that was dominant in early vertebrates, like the heavily armored placoderm fishes. In humans, our exoskeleton is expressed as our mouth full of teeth. One of the dominant trends in vertebrate evolution has been the decline in production of exoskeleton, at the expense of endoskeleton, and this is exemplified in the model of tetrapod digit evolution espoused by Sordino and his colleagues. But before examining this model it is important to understand the past interpretations of the role of heterochrony in the evolution of the tetrapod limb. For the sharing of a common limb plan is the link between all tetrapods. This limb plan consists of a proximal bone (called a humerus or a femur, depending on whether we are dealing with an arm or a leg), articulating with a pair of distal bones (radius-ulna in the arm, tibia-fibula in the leg), these articulating with carpels and tarsals (the bones of the hand and foot, respectively), which finally connect with the digits. As Shubin and Alberch point out: "Organisms share a similar pattern because of commonality of descent"—that is, they share a common ancestor.

The study of ontogenetic development is very important in determining homology—in other words, we must be confident, for example, that what we call the humerus in our arm and in the arm of an amphibian are really developmentally equivalent structures. Unless we can do this, any attempts to interpret evolutionary relationships will be in vain. One of the classic concepts of embryology, formulated in the 1820s by the great German embryologist von Baer, is that the quest for ancestral relationships should be sought in early developmental patterns. The closer these patterns, the closer the relationship between the organisms. This concept has been used by one school of thought to interpret the relationships between different tetrapods. The alternative view was based on classic heterochronic principles, where early developmental patterns were compared with later developmental stages in other organisms. Thus, classically the early embryological patterns of development of tetrapod limbs have been compared with either adult crossopterygian or lungfish fins. In this scheme the early developmental patterns were assumed to be identical, only later ones varying by the terminal addition of new, more complex structures. Most comparative anatomists who have addressed the question of the origin of tetrapod limbs have taken a classic Haeckelian approach of recapitulation, and indeed this whole scenario of fish fin to amphibian limb to reptile and mammal limb was the bulwark of the recapitulationists' creed.

While a few anatomists have argued that limb evolution has proceeded from the simple to the more complex, in terms of numbers of skeletal elements, more have interpreted it as a simplification, tetrapods having fewer skeletal elements than their fishy ancestors (even though the bones that have remained are larger and more complex). Even so, the tendency was to interpret this in terms of recapitulation. Recent work, focusing in particular on the detailed embryological development of the tetrapod limb and on the underlying genetic control, has discredited this simplistic view. But the proverbial baby should not be thrown out with the bathwater—the underlying control is still by heterochrony. The patterns, however, are just that much more complex. What some of this recent research, particularly by Richard Hinchliffe of the University College of Wales at Aberystwyth, on the comparative development of limbs in salamanders, frogs, chicks, and mice has shown is that development does not consist of a progressive fusion and reduction, or addition of new elements, like some sort of biological Lego construction kit. Rather, it arises simply by differential growth and specialization of individual elements, each following its own ontogenetic destiny. Thus, for instance, peramorphic increase in size and complexity of the shoulder girdle, humerus/femur, and radius-ulna/tibia-fibula was critical to the tetrapods' conquest of the land; as was development of the arm bones in birds, which allowed some small, early dinosaur to escape into the air with the power of flight. So, too, the independent conquest of the air by mammals as digits of the hand likewise underwent an explosive peramorphic growth in bats, allowing a body-supporting membrane to develop.

Three basic patterns of growth are involved in limb development. In the first, elements appear *de novo*, for example, the appearance of the femur or humerus in early vertebrate evolution. Second, an existing, simple element may experience Y-shaped branching into two— for example, the radius and ulna forming out of the humerus. Third, a single element gives rise to single distal condensations, either by budding of a new cartilage condensation from a preexisting element or from a single precartilage rod breaking into two separate elements. According to Shubin and Alberch, "the formation of the limb skeleton is the result of a process of sequential determination and differentiation of the various cartilage elements"—in other words, sequential formation of the parts, and their subsequent differential growth. Again, variations in the relative timing of bifurcations or induction of the elements or in the rate at which each grows relative to its ancestral equivalent and to its partners, and to the relative time that growth ceases, allow for a multitude of possible designs to evolve. Whether any one of these particular evolutionary cocktails would prove to be successful then depends on its utility, function, and how it allows a new ecological niche to be filled. Consequently, it is not possible to talk about a particular limb being "paedomorphic" or "peramorphic," in the same way that in most cases, species should not be spoken of as being "paedomorphic" or "peramorphic." In virtually all instances the limb will be both paedomorphic *and* peramorphic. The variable factor will be the ingredients in the developmental cocktail. In the evolution of tetrapods, there is a general trend toward a paedomorphic reduction in the number of elements, but a peramorphic increase in relative shape and size of those elements that remain. Such a pattern appears to be a relatively frequent event in evolution and, along with miniaturization arising from progenesis, appears to be a major process in macroevolution.

Evolution of Turtles

The evolution of turtles is a classic case of how a mix of paedomorphic and peramorphic characters, if produced in the right combination, has the power to open up new evolutionary pathways. Mike Lee of the University of Sydney has proposed that turtles evolved from what are arguably one of evolution's most ugly group of creatures, Permian reptiles called pareiasaurs.[13] Through a series of

 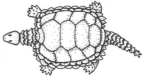

Proganochelys 210 million years

 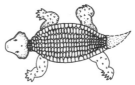

Anthodon 248 million years

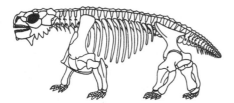 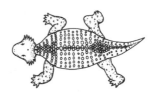

Scutosaurus 248 million years

Bradysaurus 255 million years

Captorhinus 260 million years

Suggested evolution of turtles from pareiasaurs. Based on drawings by Mike Lee.

intermediate pareiasaurs over a 50-million-year period, the earliest "true" turtle, *Proganochelys*, which lived about 210 million years ago, evolved a smaller head, an increased number of neck vertebrae (from five to eight), but a decreased number of back vertebrae (from fourteen to ten). However, these vertebrae are relatively larger—so again we see fewer structures, but larger. The changes in number of neck and back vertebrae, like the changes in segment number in arthropods, arose from alterations in the number of basic segments laid down early in embryological development, under the control of homeotic genes (see Chap. 3).

The evolution of the typical turtle shell is, Lee believes, a classic case of a structure changing its function over time. Early pareiasaurs, such as the 255-million-year-old *Bradysaurus*, had a single row of unfused bony plates running along their backs. In later pareiasaurs, these plates had spread across the entire back of the animal and increased in size. Late pareiasaurs, like the 248-million-year-old *Anthodon*, possessed an interlocking mosaic of plates. The transition from this to *Proganochelys* was relatively simple, the individual plates fusing to become a solid shield. Originally evolved as a supporting row of bony plates, their fusion into a shield covering the entire body resulted in a change in function, to one of protection. Like many other groups of animals, turtles are a mosaic of heterochronic characters: increases in numbers of some elements, decreases in others; increases in size and complexity of some anatomical features, reduction and simplification in others.

Pointing the Finger

Another significant but largely unrecognized evolutionary process is what I have called *sequential heterochrony*. This, I believe, is an especially important macroevolutionary process that allows radical morphologies to develop (geologically speaking) rapidly, but with minimal genetic disturbance to the organism. Sordino and colleagues' model for the evolution of digits in tetrapods is one such example. In their examination of the embryological development of the fin of the teleost zebrafish *Danio rerio*, Sordino and his colleagues found that the first condensation of cells of the fin bud is formed by a thickening and growth of patches of mesodermal cells. These are surrounded by an ectoderm layer, much the same as occurs in tetrapod limb buds. However, very rapidly ectoderm protrudes and folds on itself. As this fold moves distally, dermal skeleton appears inside the fold. This coincides with a sudden reduction in production of mesenchymal cells. This time of transition determines the relative extent of endodermal, compared with ectodermal, skeleton; and it is from ectodermal skeleton that the fin rays are constructed.

In the pelvic fin of the zebrafish this transition occurs developmentally very quickly, so that there is little, if any, endoskeleton; only exoskeleton forms the fin rays. However, in tetrapods this transition simply does not occur, and the limb develops by production entirely of endoskeleton, each element being sequentially formed, until finally, continued proliferation of endodermal cells produces digits as the terminal expression of the endoskeleton. Thus, digits cannot be considered as the homologues of fin rays. In the intermediate condition, found in lobe-finned fishes such as *Eusthenopteron*, there is an intermediate transition time from endoderm to ectoderm expression. Compared with the ancestral state in fishes that possess only fins, in lobe-finned fishes there has been a delay in the timing of initiation of the folding of the ectoderm, allowing sufficient time for more endoskeleton production, and thus the generation of some of the skeletal elements that occur in tetrapods. However, the late onset of ectoderm production causes the fin rays to sprout from these endodermal bones: an intermediate transition time produces an intermediate morphology.

Of particular interest at present to many developmental biologists are the factors that determine the overall morphology of the limb, especially the number and disposition of the individual skeletal elements in the limb. Signaling agents, particularly retinoids (see Chap. 3), are important in controlling the relative timing of the initiation of growth of individual skeletal elements, along with the vertebrate protein known as *hedgehog*. This protein determines skeletal element identity across the limb, and is the factor that establishes whether a digit forms as a finger or as a thumb, for instance. Fibroblast growth factors generate the signal that makes the limb bud elongate and orchestrate the sequential formation of bones down the limb. Variations in the timing of production of these growth factors are likely to affect the number and size of the bones.

Cell differentiation is controlled by a series of signals generated by proteins known as *transforming growth factors*. Research on limb development in mice indicates that a bone morphogenetic protein is produced before the transforming growth factor. It has been suggested recently that different types of bone morphogenetic proteins are active in different parts of the skeleton. In terms of heterochrony, failure of a particular protein during development can result in reduction in the size of the limbs. One particular such mutation, called *brachypodism*, results in mice growing with minute limbs and tiny paws, as well as fewer bones in the digits than normal. This occurs because the mutation produces a reduction in the number of founder cells for a particular skeletal element, so causing shortening of the bone. The unraveling by developmental biologists of the molecular basis for skeletal growth will go a long way toward helping to explain exactly how anatomical changes occur during evolution.

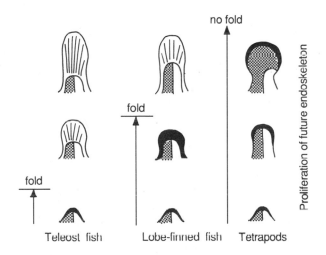

Proposed evolution of digits in tetrapods. Progressive delay in the transition from dermal to endoskeletal production arising from a progressive delay in folding of the ectoderm caused generation of digits rather than fins. In teleost fish, very early folding of the ectoderm allowed extensive dermal skeleton growth and fin rays to form. In tetrapods, the folding never occurred, so only endoskeleton was formed, producing digits. In lobe-finned fishes, the intermediate situation meant that in the pectoral appendage, for instance, the humerus, ulna, and radius formed, but from these sprouted fin rays after folding of the ectoderm.

Delays in transition from one developmental stage to another resulting in a stretching out of the growth stages probably have their underlying cause in delays in the production of particular growth factors of other proteins that orchestrate the normal course of development. I originally formulated this concept not on the basis of vertebrates but on studies that I carried out in the early 1980s on fossil trilobites from Middle Cambrian rocks (about 520 million years old) from Queensland in Australia.[14] Undertaking a study of the occurrence of paedomorphosis in xystridurid trilobites, I was puzzled by the fact that one small genus, *Galahetes*, clearly showed a range of paedomorphic features, compared with the usual species of *Xystridura*, being very reminiscent in a number of features of juvenile stages of *Xystridura*. It was also the smallest of the xystridurids, leading me to assume that it was a progenetic form; in other words, early onset of maturity had truncated its growth. Progenesis like this I had recorded in a number of other Cambrian trilobites, including the olenellid trilobites from Scotland that had got me so intrigued with heterochrony in the first place. The big difference with *Galahetes*, however, was the fact that unlike all other presumed progenetic trilobites, it had its full complement of body segments. Now, when trilobites grew from an egg, being arthropods they grew by passing through a series of molts, progressively adding body segments until maturity was reached. So in the Scottish progenetic trilobites the smaller, paedomorphic species had fewer body segments, as would be expected if there was premature termination of growth. But *Galahetes* didn't follow this nice pattern. My explanation for the fact that it was small and passed through all of its molts to reach its ancestral number of segments, but was paedomorphic in most other respects, was that rather than just a simple early termination of growth, each intermolt period was prematurely truncated. As a shorter period of time was spent between molts, there was less morphological change. This alternative method of producing paedomorphosis I call *sequential progenesis*, implying precocious onset of molting to the next stage, rather than precocious onset of just the final growth stage. This final transition from the last juvenile to the adult stage I call *terminal progenesis*. The corollary to this would be the existence of *sequential hypermorphosis*, where the various preadult developmental stages were prolonged, and *terminal hypermorphosis*, where only the juvenile to adult transition was de-

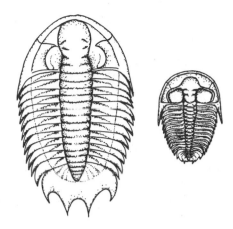

The Middle Cambrian trilobites *Xystridura* (*left*) and *Galahetes* (*right*). While the smaller *Galahetes* shows many paedomorphic traits, it has the same number of body segments as *Xystridura*. This suggests that it was a product of sequential progenesis, the time between molts and formation of segments being reduced.

layed. As it has since turned out, it looks as though these two hypermorphic phenomena are as common, if not more so, than their progenetic equivalents. And where both sequential and terminal hypermorphosis combine, the macroevolutionary effects can be profound, as I shall show in the last chapter.

While the effect of terminal hypermorphosis will be to produce a peramorphic descendant, in other words, its growth will be extended, allowing it to become larger and, in many respects, morphologically more complex (see Chap. 9), sequential hypermorphosis would seem, paradoxically, to produce some concomitant paedomorphic effects. The lack of recognition of this phenomenon has, to say the least, caused a fair degree of confusion in the interpretation of the heterochronic mechanisms that led to the evolution of a number of organisms, especially in humans (see Chap. 12). While sequential hypermorphosis describes a delay in the offset of growth of one phase, it could also be viewed as a delayed *onset* of the next phase (postdisplacement, in heterochronic parlance). Viewed like this the paradox evaporates, because postdisplacement is one of the three paedomorphic processes. So, in the case of tetrapod limbs, the delay in onset of transition from endodermal to ectodermal skeleton production means that in the lobe-finned fishes, the earlier developmental state of the limb continues into a later juvenile developmental period, but in the tetrapod this ancestral early juvenile feature carries over into the adult—in other words, classic paedomorphosis. However, in terms of the extent of development of the endodermal skeletal phase, it has a longer period of growth, generating increased complexity and resulting, furthermore, in the condensation of novel skeletal elements.

The rich diversity of life that has evolved into virtually every ecological nook and cranny on this planet in half a billion years is a product of these developmental trade-offs. Most organisms are a kaleidoscope of parts that, compared with their ancestors, have undergone greater ontogenetic development, or less. As a result, parts of the anatomy may be more "complex," others in the same body more "simple," having developed to a lesser extent. Some parts may be more numerous, while others may be less so. With such inherent developmental plasticity, the number of potential life forms that can evolve (and obviously have done so over the vast stretches of geological time) is enormous.

Many evolutionary biologists talk about "developmental constraint" as a factor that confines the scope of a species, its morphology and behavior. But few have argued that what evolution really represents is a breakdown of these constraints. When the genetic program regulating the "orderly" progression of development changes, in some way or other—a little more growth hormone here, a little less there; a slight tinkering with the time of maturation or the transition of one growth phase to another—herein lies the key to biological diversity. Developmental give-and-take generates profound morphological novelties that can open up entirely new evolutionary vistas—the tetrapod limb, which allowed vertebrate colonization of the land, was perhaps one of the greatest.

9

Evolving the Shapes Beyond

Why should dogs be senile at fourteen
and parrots sprightly at a hundred? Why
should female humans become sterile
in the forties, while female crocodiles
continue to lay eggs into their third
century? Why in heaven's name should
a pike live to two hundred without
showing any signs of senility?

Aldous Huxley, *After Many a Summer*
Dies the Swan

WHEN VIEWED OVER THE UNIMAGINABLE IMMENSITY OF GEOLOGICAL time, measured in billions of years, the biosphere as a whole can be seen to have achieved progressively larger body sizes. This increase has not been at the expense of the smaller sizes, but instead reflects an overall increase in the diversity of body sizes. The earliest life forms, the minute prokaryotic bacteria whose traces can still be found in some unaltered rocks 3.5 billion years old, can be measured in microns (i.e., thousandths of a millimeter). With the first eukaryotes, some 1.5 billion years later, cell size, as well as cell complexity, increased a thousandfold to relatively "giant" protists at least a millimeter in length. Another one billion years later, with cells getting together to form the first multicellular animals, the soft-bodied Ediacaran fauna of vaguely jellyfishlike and wormlike creatures, body sizes again increased another thousandfold. During the last half a billion years, when the animal and plant life on Earth exploded into its immensely rich diversity, body size increase slowed down a little, to a modest hundredfold, reaching its acme in the hundred-meter-high Giant Sequoia. But all the while, bacteria-sized organisms, much like their three-and-a-half-billion-year-old ancestors, can be found lurking in every corner of the Earth.

How to Get Big— Really Big

There is little doubt that the amazing appeal of dinosaurs to children (as well as to a few of us paedomorphic paleontologists) is the fact that they were big—really big. Had most dinosaurs not advanced beyond the size of a chicken, then it is most unlikely that dinosaurmania would have gripped the world to the extent that it has. Although, as I know from personal experience, a hungry chicken can impart quite a nasty peck, the scene in the movie *Jurassic Park* of the children being trapped in the kitchen by huge, rampant *Velociraptors* would surely have lost a lot of its gripping intensity had these beasts been no bigger than a rooster. Who, then, would have been chasing whom, I wonder? We all know that dinosaurs, or at least a lot of them, were big—but how did they achieve this growth? Surprisingly little research has been undertaken on evolutionary trends in dinosaurs to establish the underlying cause of this large body size. Many different lineages of dinosaurs underwent a great increase in size, paralleling patterns revealed in the fossil record by a great range of different types of marine and terrestrial animals. For instance, Tony Hallam, from the University of Birmingham in England, found that almost all of the lineages of Jurassic bivalves and ammonites that he examined (56 in all) showed that adult body sizes increased in younger species.[1] Likewise, many lineages of foraminifers predominantly show size increases. In University of Tennessee paleontologist Mike McKinney's compilation of body size trends in a range of organisms, from primates to protozoans and mammals to mollusks, 95 had undergone a size increase, only 30 a decrease, and of the latter 19 were in one group—Pleistocene mammals. The seeming preponderance of size increase over geological time revealed by the fossil record has been called "Cope's rule," after the nineteenth-century American paleontologist Edward Drinker Cope.

If we look on the grand scale from a bacterium to an elephant in terms of individual life histories, then in general terms the larger the organism, the longer its life span. So, while the generation time of some bacteria just microns long can be measured in minutes, generation time of trees nearly one hundred meters long can be measured in terms of hundreds of years. Likewise with animals: a housefly might be lucky to see out a month, but an elephant can be expected to live for up to about seventy-five years. Put basically, the larger the animal, the greater the number of cells that have to be produced. More cells can be gained either by accelerating growth rates, so that more are produced in a given time (the heterochronic process of acceleration) or by giving the organism a longer period of active growth (hypermorphosis). Most organisms have finite growth—that is to say, they have an early growth spurt, then reach an optimum size at which they remain, by and large, until they die. Growth rates are usually highest earlier in development as growth hormones actively circulate throughout the body, reduce a little as juvenile development proceeds, then decline to about zero at onset of maturity. As sex hormones are produced, growth hormones go into decline (although exceptions to this are seen in the adolescent growth spurts in one particular primate species, *Homo sapiens*).

Cell size can also influence body size. By inferring the cell size of species of the fruit fly *Drosophila* from the spacing of hairs on the wing (one hair per cell), size of facets of the eye (one photoreceptor cell per lens), or spacing of the bristles on the feet (every other cell carrying a bristle), it has been shown that small flies 2 mm long have cells 15 μm in diameter, but "giants" 8 mm long have cells 25 μm across.[2] Obviously there are limits to how far you can take this. While elephants certainly do not have cells measured in centimeters, within closely related species, such as these fruit flies, such cell size differences may manifest themselves in overall unequal body size.

Dinosaur Growth

I have talked in earlier chapters about how proportions of the body change relative to each other during these growth phases, the so-called allometric changes. In humans the head becomes relatively smaller, compared with body size (negative allometry), whereas leg length shows a relative increase (positive allometry). The consequence of any extension of the preadult growth period, by a delay in the onset of maturity, can have a profound impact, not only on the body sizes that can be achieved, but also on the overall shapes and sizes of different *parts* of the body. And dinosaurs provide one of the better examples of the effect that these increases in body size had on other aspects of their anatomy, some of the more curious features, such as what on earth did *Tyrannosaurus rex* use its pathetic little fore-limbs for? And how could *Triceratops* end up with such a head full of horns?

Until the last decade, most of what we know about dinosaurs— their great diversity of shapes and sizes, their behavior, and their evo-lutionary relationships—has been gleaned from a relatively small number of complete specimens and a lot of fragmentary material. Al-most without exception, the fossils that were used to reconstruct these ancient reptilian King Kongs were of adult specimens. Conse-quently, how we perceive the evolutionary relationships of dinosaurs has been colored, to a large extent, by the morphological characteris-tics of these large adults. By ignoring developmental data, the under-lying evolutionary processes that led to the generation of the great di-versity of dinosaur shapes and sizes have been poorly understood. However, during the last decade there has been an upsurge in interest in the developmental history of dinosaurs, fueled, to a large degree, by discoveries of eggs, which have occasionally yielded exquisitely preserved embryos, as well as early juvenile dinosaurs from a number of localities, principally in North America. As this preadult material gets described, so it becomes possible to reconstruct a picture of the ontogenetic development of many groups of dinosaurs. And from this we can start to understand the major trends in their evolution.

The largest land predators that evolution has ever thrown up are theropod dinosaurs. Recently described fossilized remains from Pata-gonia demonstrate that 100 million years ago there lived one of these dinosaurs, recently named *Giganotosaurus*, that was about 12.5 m long and weighed 6 to 8 tons.[3] Such huge theropods, and the more familiar tyrannosaurids, possessed several derived features that probably arose from heterochrony.[4] Particularly striking, apart from their massive body size, is the large head relative to body size. Yet when we compare the skull of a huge adult with that of a delicate posthatchling thero-pod, one of the striking things is how the skulls of the juveniles were small and elegant, relative to the body as a whole. Here we have a feature that is common in many lineages where body size has increased—allometric scaling of the feature with the large body size. In other words, the comparative extent of growth of the skull compared with the rest of the body was greater. Thus the larger the body, the relatively larger the skull gets. However, even within the skull, allometries varied—the different bones that go to make up the skull and the lower jaw did not all grow at the same degree, relative to each other. Growth was greater dorso-ventrally than antero-posteriorly, with the result that the head changed shape during ontogeny, from being relatively long and slender to becoming more massive and foreshortened ("morphing" was clearly not an invention of late-twentieth-century computer programmers).

The same peramorphic effect of allometric scaling that dictated the skull shape in *Tyrannosaurus* probably explains the evolution of the massive hind limbs in tyrannosaurids, particularly in the larger forms. However, the characteristic very short forelimbs and hands, which carry just two fingers, are paedomorphic features. People have

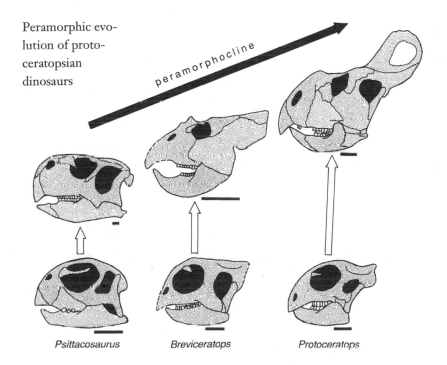

Peramorphic evolution of protoceratopsian dinosaurs

peramorphocline

Psittacosaurus   Breviceratops   Protoceratops

argued long and hard about how these arms were used, trying to explain just what their functional significance could be. Were they used to lever the beast up when it fell over, as some have suggested? Could they have been grapples used during mating to stop the amorous pair from tripping over one another during the height of their passion? In all likelihood, explanations like these were wrong. *Tyrannosaurus* probably couldn't even have scratched its nose with these pathetic limbs, for the simple reason that they had no functional use whatsoever. You and I have an appendix. It's more bother than it's worth. Likewise, most flightless birds are encumbered with wings that serve no useful purpose whatsoever. Penguins are one of the few exceptions. Such structures are mere evolutionary baggage, the flotsam and jetsam of evolution that has lost out in the great developmental trade-off. So it probably was with the arms of *Tyrannosaurus*.

If selection favored the tyrannosaurid with a massive body; huge, powerful legs that could allow it to outrun its potential dinner; and a head, neck, muscles, and teeth able to rip a nice, tasty hadrosaur to shreds, who needs arms? Birds cope very well without arms to pick up their food; *Tyrannosaurus* was the same. As selection favored these peramorphic traits that had developed "beyond" those of ancestral theropods, then the structures that lost out in the evolutionary lottery were the arms. Almost by definition they are paedomorphic features, probably akin to the feeble arms possessed by embryonic early theropods. In this particular evolutionary cocktail, the "overdeveloped" structures were overwhelmingly the strongest flavor; the tiny arms just a drop in the glass.

Analysis of tyrannosaurid bone structure suggests that, as with many other dinosaurs, theropod growth rate was relatively rapid. Peramorphic features in tyrannosaurids, even in large forms, may therefore not have been entirely a function of delayed onset of sexual maturity but could also have been caused by acceleration in the growth of many structures. Perhaps here we see the clue to what was controlling the large body sizes attained by many different groups of dinosaurs: hypermorphic delay in maturation, combined with accelerated growth, the two forming a particularly potent mix in the Mesozoic. This idea of rapid growth is supported by the discoveries made by David Varricchio of the Museum of the Rockies in Montana. Sectioning bones of the theropod *Troodon*, Varricchio found that the young dinosaurs seem to have passed through three ontogenetic growth phases. This was demonstrated by changes that he saw in the bone microstructure. The thin sections of the bones revealed the presence of lines indicating periods of arrested growth. Varricchio thinks that these may reflect the seasonal climate in which the dinosaurs lived—growth during the seasons of plenty; little or no growth in the lean seasons. If this is so, it may give us a handle on actual growth rates and allow us to establish the time of onset of sexual maturity. On the basis of his data, Varricchio came to the conclusion that the species of *Troodon* that he was studying may have reached maturity after three to five years, at a body weight of about 50 kg.[5]

Comparable work on bone microstructure lends support to the view that larger dinosaurs took longer to reach maturity. Anusuya Chinsamy of the South African Museum in Cape Town considers, on the basis of bone microstructure, that the 20-kg theropod *Syntarsus* reached a mature body size after seven years, whereas the much larger prosauropod *Massospondylus* took upward of 15 years. Similar analysis of bone microstructure in other dinosaurs may sometime in the future allow us to suggest the actual heterochronic mechanisms that were involved in the evolution of dinosaurs.

Interestingly, the tyrannosaurids show one of the few distinct examples of paedomorphosis in dinosaurs. By far the smallest of the tyrannosaurids were *Nanotyrannus* and *Maleevosaurus*. As a consequence of their pattern of juvenile growth, they retained as adults a much more slender skull than did their more famous relative, *T. rex*. These Late Cretaceous tyrannosaurs were "only" about 5 m in length. The effect of paedomorphosis can be seen not only in their small body size and slender snout, but also in the big, rounded eye orbit. However, even *Nanotyrannus* got into the peramorphic act in a small way. While Bob Bakker, at the University of Colorado at Boulder, has suggested "neoteny" as a way of producing the delicate adult lacrymal "horn" that projects above the eye in *Nanotyrannus*, the fact that it most likely arose from a hornless and unswollen ancestral state would seem to imply a peramorphic origin, since juvenile tyrannosaurs lack the development of this "horn."[6]

Among the herbivorous dinosaurs, hadrosaurs are characterized by the development of a wide array of startling and bizarre crests. These are a particular feature of the late-stage ontogeny. Consequently, any slight variations in the time of onset of sexual maturity and in the cessation of growth can have pronounced effects on the degree of skull crest development. The recent discoveries of juvenile and embryonic specimens of the hadrosaurid *Maiasaura* and the hypsilophodontid *Orodromeus* from western Montana by Jack Horner of the Museum of the Rockies has thrown new light on the nature of early embryonic development in both of these groups of dinosaurs. In particular, analysis by Horner and Phil Currie at the Tyrrel Museum in Drumheller, Canada, of the bone structure of embryonic and nestling skeletons of the hadrosaur *Hypacrosaurus* shows that, like the tyrannosaurids, they grew at a feverishly rapid rate. Thin sections of the bones of this tiny dinosaur show it to be highly vascularized, with a great deal of calcified cartilage, indicating very rapid, early growth.[7]

By looking at the ontogenetic changes in the hypsilophodontid *Dryosaurus*, it can be seen that the long snout that develops in adults is a peramorphic feature arising by a relative increase in growth of the nasal and frontal bones of the skull, as is the case in hadrosaurs. *Tenontosaurus* is even more peramorphic than *Dryosaurus*, the adults having a longer snout, smaller orbit (these being relatively larger in the juveniles), and absence of premaxillary teeth. As with the tyrannosaurids, reduction and modification of fingers and reduced forearm size occurred in *Tenontosaurus*. The increase in growth to reach a large body size (up to 7.5 m) in *Tenontosaurus* (hypsilophodontids are usually less than half this size) may explain the reduction in size of the fingers and forearms, these elements being dissociated from the overall peramorphic trends in cranial characters as body size increased.

Recent studies of some of the early ceratopsian dinosaurs (the group that includes *Triceratops* as an end-member) have revealed a range of modifications in body proportions and growth of structures during ontogeny. By examining the relative extent of growth of these features, again the overwhelming trend is of peramorphosis, as successive species went through more and more growth during ontogeny. When we look at the skull of very early, primitive ceratopsians, such as *Psittacosaurus*, we can see that as the skull grew from just a couple of centimeters long to about 20 cm long, the changes were not very pronounced. The orbit, in which the eye sat, became relatively a little smaller, and the nasal region became slightly beaklike. But as we track up through successive ceratopsians, we can see that in addition to an increase in body size, the skull went through increasingly more changes. Whereas the hatchling *Breviceratops* looked remarkably like the older *Psittacosaurus*, adult *Breviceratops* had a skull some 30 cm long, which featured a pronounced beak, smaller orbits, and increased growth of the back portion to produce a pronounced frill. These trends were accentuated even further in *Protoceratops* and finally *Triceratops*. Yet the tiny hatchlings of *Protoceratops* look, for all the world, like the young of their distant ancestor *Psittacosaurus*. As body size increased, the growth trends continued, and the ornate frills and horns on the later ceratopsians developed. On another day, in another age, this would have been called recapitulation; today we call it peramorphosis.

Horsing About   It is now time to revisit another icon of evolution. I showed in Chapter 6 how the evolution of "Darwin's" finches needs to be looked at from the inside as well as from the outside—from an external adaptive perspective. And so it is with horses. For many of the features that so characterize this famous example of evolution, as in many of the dinosaurs, have their origins in extended peramorphic development.

There can be few museums carrying at least a smattering of natural science displays that haven't at one time or another had a display featuring the evolution of horses. In many cases (excuse the pun), these displays have remained unchanged for a couple of decades. There it lurks, in the corner of a slightly murky case: a fading image of a "classic" evolutionary trend, showing how during the Tertiary there were just five types of horses: *Hyracotherium* (otherwise known as *Eohippus*), which evolved into *Mesohippus*, which in turn evolved into *Merychippus*, *Pliohippus*, and then *Equus*. And the caption talks about how these fossil horses demonstrate that evolution proceeded in a very nice, neat, orderly direction—that of increasing complexity and morphological sophistication in a single evolutionary trend. The illustration will purport to illustrate two main trends: one of a reduction in the number of toes on each foot, from four down to the one that supports every living horse, from the racehorse to the draft horse; and a change in the nature of the tooth, from low and simple to high and more complex. These changes will be explained from a typically adaptationist viewpoint, as being adaptations that evolved as the horses moved from living in woodlands, where they browsed on the leaves of trees, to galloping across the grasslands, fueled by a diet of tough grass.

Such an explanation is misleading from many points of view, not the least of which is the gross oversimplification and the assumption that the evolution of all horses was purely unidirectional—it wasn't. But thanks to the efforts of a number of vertebrate paleontologists, principal among these in recent times being Bruce MacFadden from the Florida State Museum, our understanding of the evolutionary radiation of horses has revealed a pattern that is exceedingly more complex, and one that is dependent as much upon the factors that underlay the anatomical changes as on the adaptive significance itself. It is not only the changing degrees of ontogenetic development that may have influenced horse evolution, but the effect of increases in body size can also be shown to have played a crucial role.[8]

The first horses that skittered through the early Eocene forests of North America, 55 million years ago, were no larger than a dog, growing to an estimated maximum body weight of around 25 kg. These early horses are distinctive in having had four toes on their forelegs and three on their hind legs. Over the subsequent 30 million years, until the early Miocene, the ten species of horses that evolved only doubled in weight. This is thought to have been a single lineage showing not only quite a modest increase in body size, but also an increase in height of the teeth. Moreover, by the Oligocene, the fourth toe on the front foot had been lost. Then around 25 million years ago an explosive evolutionary radiation occurred, generating a diverse assemblage of horses—some larger, some smaller, and some the same size as the Oligocene horses. Most, it is true, grew much larger than their pre-Miocene ancestors, many exceeding 100 kg in weight. Some even reached 400 kg by 10 million years ago. Yet despite this evolutionary size expansion, some smaller forms as light as 70 kg still existed. As body size increased in some lineages, so too did the height of the teeth, enabling them to withstand the more abrasive grasses that were appearing while the continent began drying out as the globe slid into the last great Ice Age.

Whereas the Miocene horses ran around on feet with three toes, later, larger horses suffered a further reduction in the size of the outer toes. But the central toe, which was to become the hoof in modern

Three stages in the evolution of horses. At the bottom, the fox-sized, 55-million-year-old *Hyracotherium;* in the center, the 20-million-year-old *Miohippus;* at the top, the living *Equus.*

horses, increased in size to support the larger body weight. By the Pleistocene, all horses were graced with a single, enlarged medial toe that we call a hoof, and body weight had soared to half a ton. Once again, evolution had hedged its bets with a developmental trade-off: a paedomorphic *reduction* in the number of toes, but a peramorphic *increase* in size of the one remaining toe.

Bruce MacFadden analyzed many of the evolutionary trends in body size between species pairs throughout the entire 55-million-year evolutionary history of horse evolution. What he found was that of 24 lineages that could be identified, 19 did, indeed, demonstrate an increase in body size. Correlating body size with length of life, he ar-

Horses showed little size change for nearly 40 million years, but for the last 20 million years of their evolution there was a great increase in diversity of body size.

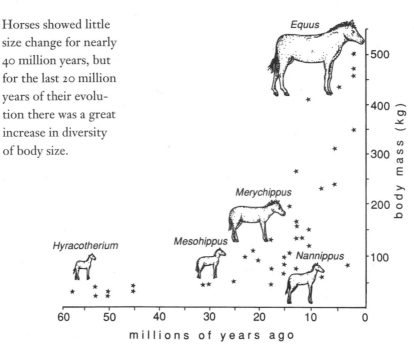

gued that the evolution of a longer growth period (hypermorphosis) allowed not just the body size to increase, but carried along with it the increase in tooth length and in hoof size. There is even evidence to suggest that some of the smaller Miocene horses attained sexual maturity earlier than later equids, providing firm evidence for the operation of hypermorphosis in later horses. Thus, delay in onset of maturity in some of the later lineages allowed a larger body size to be attained, and this in turn dragged along proportionate changes in certain structures, such as the form of the teeth and legs. Dissociation of the side toes saw their paedomorphic reduction and ultimate loss. However, evidence that the genetic signal for the production of side toes still lurks deep within the equine genome is revealed by the occasional throwback that sprouts the odd pair of small extra toes.

Why Grow Bigger?

One of the more commonly held beliefs in biology is that animals attain a large size simply by growing for a longer period (in other words, hypermorphosis). On a broad scale, such a "mouse to elephant" type of comparison is undoubtedly true. However, within some lineages, it may be that this is not the only mechanism for size increase. Brian Shea from Northwestern University has shown how, in some primates, closely related species often attain a larger body size just by growing faster (acceleration) for the same period of time as the smaller species.[9] A study of the times of maturation in comparison with body size in carnivores carried out by John Gittleman of the University of Tennessee at Knoxville revealed that larger forms very often have such a faster growth rate. However, it is often in conjunction with a hypermorphic extension of the growth period.[10] Such was probably the case in dinosaur evolution.

There are clear survival advantages to attaining a large body size quicker by acceleration. One in particular is that it takes prey species out of the range of optimum prey size quicker. If, as a dinosaur, you grow slowly after you have hatched from your shell, and so stay smaller for a longer period of time, then the chances of your being picked up in the jaws of a hungry predator are that bit higher than if you had grown faster, gotten bigger, and increased your chances of survival. The very high juvenile growth rates that dinosaurs appear to have experienced are likely to have come about as a response to high predation pressure. Although, as a young tyrannosaurid, you might have been penciled in on some fellow dinosaur's menu, at the other end of the scale, by getting as large as possible in as short a period of time as possible, you would have been able to turn the tables. The larger you got, the better your chances of avoiding being eaten yourself, but the better your chances of becoming a successful predator and surviving long enough to mate and pass on your genes to another generation. The downside of this, and the factor that most effectively puts the brakes on your accelerating growth rate and limits your body size, is your metabolism. If you grow too large, then your energy input has to be that much greater. Large animals get around this problem to a certain degree by becoming decidedly less fussy about what they eat, feeding on a wider range of items than smaller species (a case in point being the primate *Homo sapiens*, which has a spectacularly broad diet, compared with, say, a lemur). With ever-increasing body size along an evolutionary lineage, a point will be reached when energy output exceeds energy input: this is the real dampener on increased body size growth.

There are other advantages to extending the juvenile phase of rapid growth. As I discuss in some detail in Chapter 12, prolongation of the juvenile growth phase in *Homo sapiens* (the second largest primate) allows an extended period of learning, arguably the single most important factor in the species' evolutionary success. Another advantage of two closely related species differing appreciably in body size is that it engenders habitat separation. Robert Martin of Berry College in Georgia has shown how one of the main advantages of the cotton rat *Sigmodon* evolving a larger size is that it produces a behavioral change: increased aggression. The effect of this is that two species of different size find it rather unpleasant to coexist in the same microhabitat. If they did, the large cotton rats would kill the smaller species that transgress into their habitat. While, in the short term, this might have the effect of driving the smaller form into another habitat, it could ultimately lead to its extinction. Examination of long-term evolutionary trends in the cotton rat reveals a predominance of body size increase.[11]

The ubiquity of Cope's rule in evolution suggests that increased size must be advantageous to the species in some way, and clearly cannot be maladaptive. In many cases, it may well be that larger body size (or the associated behavioral or life history factors—see Chap. 10) are the principal targets of selection, morphological traits being dragged along and contributing, in some manner, to the "fitness" of the species. However, a trait that is an evolutionary success at one time may prove to be a species' nemesis at another, becoming maladaptive if there is a major environmental perturbation. Body size is particularly susceptible to the vagaries of environmental fluctuations. As I have mentioned, the price to pay for large body size is the requirement for either a large energy input or a decrease in metabolic rate to minimize energy output. Nevertheless, a body size and metabolic equilibrium established for a particular set of environmental conditions may prove inadequate during periods of environmental degradation. Basically, larger animals require more food than smaller ones. Consequently, when the environmental going gets tough, it's the large species that suffer first. Striking examples of this are the extinctions of the dinosaurs at the end of the Cretaceous (dinosaurs in the 1–10 ton weight range being at their acme at the end of the Cretaceous) and the catastrophic loss of the vertebrate terrestrial megafauna during the Late Pleistocene in a number of continents, induced by a combination of major climatic deterioration and the arrival of a major new predator in the form of *Homo sapiens*.

Relatively minor environmental fluctuations can also have an impact on body size. One of these is temperature. Codified as "Bergmann's rule," larger body size in warm-blooded animals seems to correlate positively with lower temperatures. However, cold-blooded organisms show the opposite relationship: their size increases in warmer climates, because they retain body heat for longer, so they stay active longer. This relationship has been used as an argument for the larger body sizes attained by dinosaurs in the Mesozoic. Global temperatures are thought to have been higher at this time, as were CO_2 levels, which reached more than five times the present-day levels, some 120 million years ago. Demonstration of Bergmann's rule over relatively short geological time scales has been provided by Bjørn Kurtén in a study of the evolution of body size in the brown bear *Ursus arctos* over the last half a million years. During each of the major glacial advances, as a consequence of reduced global temperatures, body size increased, declining in the intervening interglacial periods. Interestingly, woolly mammoths from the same time period underwent size increase during the warmer interstadials of the Late Pleistocene, whereas the "Irish elk," *Megaloceros*, seems to have suffered no effect at all from temperature fluctuations. In all probability, body size changes are influenced by a complex interaction of environmental factors.

On a larger time scale, it is possible that environmental factors other than temperature may have contributed to increased body size. One group that doesn't seem to show a general increase in body size over time is the insects. Their period of largest size occurred not very long after their evolution in the Early Devonian, about 400 million years ago (see Chap. 3), when the giants of insect evolution, such as dragonflies that had a wingspan of up to 70 cm, evolved during the Carboniferous, some 300 million years ago. Computer models suggest that at this time oxygen levels in the atmosphere increased to their highest levels (about 35 percent of atmospheric gases, compared with 21 percent today). It has been proposed that increased oxygen levels may have caused enhanced metabolisms.[12] Moreover, atmospheric pressures would have been higher, allowing larger bodies to be carried aloft. Support for the effect of elevated oxygen levels influencing body size comes from the last of the pterodactyls. The giant *Quetzalcoatlus*, which had an estimated wingspan of 12 meters, lived during the Late Cretaceous. This was also a time of elevated oxygen levels.

On the Horns of a Dilemma

Two of the most well-known examples of peramorphic evolution in the fossil record involve a combination of increased body size and a spectacular increase in horn size. Roaming over much of Europe and western Asia during the Pleistocene, from about 450,000 to 11,000 years ago, was the largest deer (cervid) that has ever lived. Called the Irish elk, on account of the frequency with which the skulls and antlers of these magnificent beasts have been dragged from the bogs in Ireland, *Megaloceros giganteus* was neither exclusively Irish nor was it an elk. It was, basically, a very large red deer. This giant deer grew to 1.8 m at the shoulder and sported antlers up to 3.5 m across.[13] For a long time its extinction at the close of the last Ice Age was attributed to the maladaptive nature of its huge antlers. These, it was argued, would have got snagged in trees, having grown too big for the good of the animal. Other colorful explanations have been offered for the extinction of *Megaloceros*. In 1697, Thomas Molyneux suggested that the extinction of the "Irish elk" in Ireland was caused by an "epidemick of distemper," which had resulted from "a certain ill constitution of the air." Archdeacon Maunsell thought, in 1825, that "they must have been destroyed by some overwhelming deluge." One Dr. MacCulloch even believed that the fossils were found standing erect, noses elevated—the deer's final gesture of defiance to the rising Flood. Yet as early as 1830 the hand of man had been suspected, one observer finding it "questionable, if the human race has not occasionally proved as formidable as a pestilence in exterminating from various districts, whole races of wild animals."

Yet far from being an evolutionary failure because of its huge antlers, *Megaloceros* enjoyed wide evolutionary success for nearly half a million years. The fact that the species evolved at all means that even if the antlers conferred no huge adaptive success on the animal, they were not maladaptive—they were, at the least, just nonadaptive. Both body and antler size remained reasonably constant over the almost half-million-year evolutionary history of the species. The only changes that occurred, according to Adrian Lister, were a reorientation of the antlers and an increase in tooth size.

A debate that has gone on for a long time has centered on determining the main targets of selection in the evolution of such an animal. Was it body size or antler size or both? A nonadaptive trait is one that appears to confer no functional advantage on the organism. Neither, however, does it confer any sort of disadvantage. In the case of *Megaloceros*, Stephen J. Gould has argued that the principal target of selection was body size. As selection favored a larger body size over time (although little is known of its presumably smaller ancestor), any structure that grew with positive allometry would have been proportionately larger. In such a scenario, the evolution of such enormous antlers would be considered to have been just an ancillary effect of its positive allometry, dragged along by the extended growth period. Presumably because the species existed with a fair degree of success over a wide geographical distribution, the antlers would be viewed as nonadaptive structures. But it needs to be remembered that in such deer, the antlers were shed and regrown each year, so the energy input required to generate such structures had to be worth its while.

However, it is unlikely that the very large antlers served no useful adaptive "purpose" at all, given the large investment in energy required to produce them annually. There is a lot of evidence from living cervids that antler size plays an important role in sexual selection, either in terms of display or from actual threat or combat. Indeed, the positive allometry displayed by the antlers may, in fact, have been a major adaptive trait in terms of sexual selection. But for a cervid to grow large horns, body size is itself a necessary adjunct, both mechanically, to support the weight, but also from a metabolic point of view. Only a large animal could ingest enough nutrients to grow such large structures annually. In mating behavior in cervids, *both* larger antlers and large body size play a crucial role. The larger the body size, the stronger the animal, in general; the larger the antlers, the more attractive the male, or the more effective in actual male-male threat or combat. The two go hand in hand. Ultimately, it was unlikely to have been the huge antlers that led to the demise of *Megaloceros*, but its large body size. *Megaloceros* was just one of the many species that made up the megafauna, which was so severely affected by environmental changes during the Late Pleistocene. A decline in food

The so-called Irish elk: *Megaloceros giganteus* was neither exclusively Irish nor an elk, but a very large red deer. Drawing by Jill Ruse.

supply is more probably the cause. Certainly, in ungulates as a whole, body size variations are greatly affected by both the quality and the length of the plant growth season.

It has been assumed that larger body size in *Megaloceros*, and consequent great enlargement of the antlers, occurred by hypermorphosis: a delay in the onset of maturity. While some estimate of the age of individuals can be made on the basis of the basal circumference of the antlers, elucidating age at onset of maturity is distinctly equivocal. Moreover, nothing is known of its ancestor, whether it was smaller, larger, or the same size, or its time of maturity. Thus we are unable to say whether or not the preadult growth period was extended in *Megaloceros*. As John Gittleman has demonstrated, larger body size can also ensue from an acceleration in actual growth rate. If body size as a whole was accelerated, then positive allometry of the antlers would mean large antlers being produced at a quicker rate. Unraveling the dual roles of hypermorphosis and acceleration, or the extent to which they both contributed to the massive skulls and antlers that have adorned many an Irish baronial hall, is likely to be a difficult exercise. However, another, much older group of mammals, where a number of ancestors and descendants can be identified, offers a better chance of establishing the relative roles of different heterochronic processes in inducing peramorphosis—the titanotheres.

These large, horned, early Tertiary mammals were widespread in North America during the Eocene and Oligocene (30 to 50 million years ago). While the earliest forms like *Eotitanops* in the Early Eocene had quite small body sizes and were hornless, Oligocene genera, such as *Brontotherium*, reached nearly three meters at the shoulder. Moreover, they carried massive horns that protruded from the front of the skull. Intermediate-sized titanotheres, such as *Protitanotherium* in the Late Eocene, possessed just mere traces of horns. The trend of increasing body size and horn size and complexity was originally interpreted by A. H. Hersh in 1934 as an extension of the ancestral positive allometric relationship between horn size and body shape arising from an increase in body size, which Hersh considered to be the target of selection. However, a more recent reinterpretation by Mike McKinney and Robert Schoch showed that hypermorphosis was insufficient to produce the large horns that evolved in the later, larger Oligocene genera. The measurements that they obtained from a range of skulls indicated that horn growth also occurred by acceleration and, in the larger titanotheres, some other traits. The Oligocene forms, *Allops, Brontotherium, Menodus,* and *Brontops*, show predisplacement of the horns. In other words, compared with their Eocene ancestors, horn development began at an earlier stage in ontogeny. Thus predisplacement, acceleration, and hypermorphosis combined to produce large descendants with proportionately much larger horns in the Oligocene. Indeed, McKinney and Schoch have calculated that had this not occurred, horn length would only have been about one-third of skull length, instead of the almost half actually attained. The operation of all three peramorphic processes argues for the horns as a target for selection. The functional importance of larger horns can be assessed by considering the degree of stress that they could have withstood. While the stress that such horns could withstand is equal to force/area, increase in body size would result in increased force of impact. The most efficient way of compensating for this was to increase horn size in order to spread the load. Hypermorphosis alone would have been insufficient.[14]

Growth of the Parts

I suppose that had I been writing this book in the late nineteenth century I would be describing the evolution of *Megaloceros* as a fine example that demonstrates the ubiquitous nature of recapitulation—an apparent terminal addition to the growth phase by an extension of the juvenile phase of growth. Certainly, peramorphosis as we describe it today illustrates a "greater" degree of growth—more complexity, if you like, in the descendant compared with the ancestor. But as the titanotheres reveal, the evolution of peramorphosis is often a complex matter of the interplay of a number of different processes. The biogenetic law of Haeckel, with recapitulation as its basic tenet, looked at the organism in its entirety as displaying these "advanced" features, having evolved "beyond" the ancestor. However, today's modern synthesis of heterochrony has shown that both peramorphosis and paedomorphosis may affect different morphological features in the same organism, even a single structure, not just the whole. Different structures may therefore dissociate and undergo their own peramorphic evolutionary history, whereas others may reduce by paedomorphosis.

Heterochrony as a generator of morphological novelty and major evolutionary breakthroughs has, since de Beer's time, focused largely on paedomorphosis, particularly by progenesis, as the key process. Yet, as I have demonstrated in the previous chapter, the mix of paedomorphic and peramorphic features can equally well promote macroevolutionary novelties. So too with local peramorphosis, which affects just one key morphological feature. Even the peramorphic acceleration of a single group of bones can have a profound morphological, behavioral, and ecological effect, triggering the invasion of a new niche and opening up a new evolutionary pathway. A spectacular example of this is found in an extinct group of reptiles that lived 230 million years ago during the Triassic Period—the prolacertiforms. These animals were characterized by the possession of an incredibly long neck, the length of which makes a giraffe's neck seem stunted by comparison.

Karl Tschanz of the University of Zurich in Switzerland thinks that this group was aquatic. Tschanz has shown that during their evolution prolacertiform reptiles increased their neck length. This was achieved by a peramorphic increase in the length of each neck vertebra and also by a peramorphic increase in the number of vertebrae. The earliest forms, such as *Prolacerta*, had eight neck vertebrae. The most extreme neck length was attained by the geologically youngest prolacertiform, *Tanystropheus longobardicus*. In this beast the neck accounted for more than half the total body length but consisted of only twelve vertebrae. However, these were individually extremely long. Thus, there had been an increase in the number of vertebrae, combined with an acceleration in the growth of each skeletal element. The earliest *Tanystropheus, T. antiquus*, which had evolved from *Prolacerta*, had nine very elongated cervical vertebrae, so it looks as though neck increase first occurred by elongation of what was already there, followed by addition of new vertebrae. Even some earlier prolacertiforms, such as *Macrocnemus bassanii*, which had relatively shorter necks, show, surprisingly, higher than average allometric coefficients for neck length and a greater variability, compared with the geologically younger *T. longobardicus*. This short-necked species was only one meter long, whereas *T. longobardicus* grew up to six meters long.[15]

The Triassic (230-million-year-old) prolacertiform reptile *Tanystropheus longobardicus*, demonstrating peramorphic evolution of neck vertebrae.

The relative decrease in variability of allometric coefficients in *T. longobardicus* may be a function of its larger body size. Tschanz argued that the relatively longer neck of *T. longobardicus* may have been due, in part, to hypermorphosis, allowing the growth period to be extended. The relative reduction in allometric variability in the younger species was, in all likelihood, a functional necessity. Had the neck vertebrae in *T. longobardicus* grown at the same rate as in *M. bassanii*, and had the animal still grown to six meters in length, the neck that it would have possessed would have been absurdly long and functionally useless, and would have become an adaptive liability. The growth rates that it attained, which produced a neck that took up about half the animal's entire length, were at its adaptive limit. Beyond that the evolution of a longer neck would have been maladaptive, probably due to the poorly developed cervical musculature. So, peramorphic evolution, by two different processes operating at different periods in the group's evolutionary history, opened up to these aquatic reptiles a lifestyle unmatched by any other vertebrates at the time.

The modular construction of plants means that local growth fields, such as leaves, shoots, flowers, fruits, seeds, and roots, can undergo heterochronic changes independent of each other. However, there is also evidence to suggest that developmental changes occurring in one module can have a direct influence on other modules, by a "knock-on" effect. One such example is the evolution of seed size. There appears to be a strong correlation between seed size and habitat. Thus, for instance, the seeds of woodland herbs are generally larger than seeds of species inhabiting more open habitats. It is thought that this is an adaptation to the occupation of different regimes of aridity or shade. Large seeds are found in more shaded or dry habitats, since their larger reserves allow establishment in less than perfect conditions. Ed Guerrant of the University of Oregon has stressed how developmental pathways followed by one module, or a set of modules, may have a direct influence on others. It has long been known that the size of different plant organs, including seeds, is related to the size of the meristems from which they develop. Likewise, a close allometric relationship can exist between different sets of plant "modules." Thus, leaf length (which itself is positively correlated with plant height) is correlated with the length of the seed. Similarly, seed size has been shown to be directly correlated with its height on the stem.[16]

There is an evolutionary drawback to plants evolving particularly large seeds. Larger seed size is often associated with lower rates of germination and with significantly lower relative growth rates of seedlings.[17] Instead of being of direct adaptive significance, these factors, rather than seed size, may be the actual targets of selection. Close correlations between seed size and flower size, or leaf length or plant height indicate that some of these features are most likely being carried along piggyback fashion. While unraveling the true targets of selection is fraught with many problems, it does demonstrate the crucial importance of understanding relative growth rates and durations of growth of different modules in plants, and their interrelationships.

The concentration of peramorphic increase in one module may be at the expense of other modules as a developmental trade-off, such as occurs in animals. One of the more stunning examples of this in plants occurs in the gigantic parasitic genus *Rafflesia*, which grows in the rain forests of Sumatra and Borneo. Growth of the leaves, stem, and roots had been paedomorphically reduced to such an extent that they were almost entirely lost. This extreme paedomorphosis has occurred as a developmental trade-off, the flower growing peramorphically and reaching a diameter in excess of one meter. Studies of the flowers' growth duration and rates would establish the relative importance of hypermorphosis and acceleration in generating the large flower size. Some smaller species have flowers that reach only 20 cm in diameter, reflecting either shorter generation times or lower growth rates.

With Wings on Their Fingers

The experience of learning to fly for young bats is fraught with many problems. Apart from trying to flex their wings in a confined space along with a few thousand other youngsters, a major problem as they are developing is the rapidly changing shape of their wings. Each day their wing shape and size vary appreciably from the preceding day. Imagine trying to learn to drive a car, and each day you sit in the driver's seat the pedals have changed position, and the steering wheel is a different size; then you will appreciate the challenge faced by a young bat. Every time it tries to fly, it is faced with having to cope with a wing anatomy whose proportions and shape relative to the body are changing at a phenomenal rate. This arises from a great acceleration in the growth rate of certain fingers that support the delicate, leathery wing. The bones that go to make up the struts that support a bat's wing are of the same form, number, and relative position to each other as the bones in your arm and hand. Where they differ is just in their shape and proportions. To understand how they could have changed in their proportion so drastically, we must delve back deep into the embryo of a bat; for here lies the key to the evolution of flight in mammals.

The bones in a bat's wing consist of an extended forearm, plus extremely long fingers. Only the thumb is not elongated. When bats are born, their wings are relatively short and incapable of allowing flight, being a mere 20 percent of the adult wingspan. By the time they have taken their first, tentative forays into the air, when they are about four weeks old, the arm and finger growth have accelerated to such an extent that the wings have achieved about 60 percent of adult size. Full wing size is gained about forty to fifty days after birth. To see the enormous extent to which the arm and hand bones grow, compared with the rest of the skeleton, Rick Adams of the University of Wisconsin at Whitewater and Scott Pedersen of the Caribbean School of Medicine in Montserrat examined preserved embryos that had been stained to show up the developing cartilage and bone. Although bats spend about fifty to sixty days *in utero,* many significant events in skeletal growth don't begin until about thirty-five days following fertilization. At this time most of the skeleton is still cartilage, the only mineralization having taken place in the jaw and collarbone. The hands and arms hardly look any different from those of other mammals at a comparable stage of development, the hand being about one-third the size of the head. However, after about 40 days of gestation the fingers undergo a rapid acceleration, outpacing the rest of the skeleton. This accelerated growth continues until just prior to birth, when they exceed the length of the forearm.[18]

The earliest known bat skeleton was found in 55-million-year-old rocks in North America. This bat, *Icaronycteris,* is remarkably similar

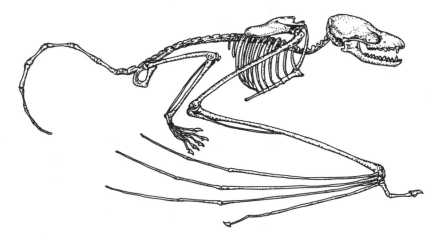

Skeleton of the earliest known bat, the 55-million-year-old *Icaronycteris.*

to living bats, the supporting wing bones being as long as those in living bats.[19] Indeed, bats were the most highly specialized of any mammal group during the Eocene and had reached the most modern grade. It has been argued that the evolution of bats may have occurred relatively rapidly, perhaps in as little as 8 million years. Echolocation, a feature possessed by living bats to detect objects in their path, appears to have been present in even these earliest bats. Evidence for this comes from the structures of the cochleas in the earliest bats. Rapid evolution of bats by acceleration of the wing-support bones may have only occurred in conjunction with the possession of echolocation in their immediate ancestors. It has recently been suggested that these ancestors were small, nocturnal gliders living in forests.[20] In addition to evolving wings, bats would have evolved an improved echolocation system, along with adaptations to minimize "self-deafening" (the ability to minimize strong signals that would otherwise mask weak signals). Localized peramorphic growth of the fingers, in combination with a membranous wing surface and inherited echolocation, allowed an aerial, insect-eating niche to be conquered.

Bats were not the first vertebrates to master the air. That honor went to flying reptiles: the pterosaurs. Although the evolution of exquisitely delicate long fingers in bats is clearly a most impressive feat of biological engineering, it tends to pale into insignificance when compared with the manner in which pterosaurs accomplished this task. Not only did flight evolve in reptiles much earlier than in mammals (during the Late Triassic, about 220 million years ago), but it was achieved biomechanically by the acceleration in growth of even fewer skeletal elements.

Flying high over the Texas landscape during the late Cretaceous, about 80 million years ago, was the largest flying creature ever to have evolved, the pterosaur *Quetzalcoatlus*. With an estimated wingspan of 12 meters and weighing about 65 kg, this creature was larger than some small airplanes. But rather than an acceleration in growth of the forearm and four fingers to support a massive wing, the skeletal element that supported the wing, as in other pterosaurs, was amazingly just the fourth finger. While the shoulder girdle has some modifications that allow attachment of powerful flight muscle, the humerus is relatively short. Neither the ulna nor the radius is exceptionally long, nor are the first three metacarpels or digits. It is all in the fourth finger. The fourth metacarpel is greatly elongated and thickened, as are the four phalanges of the finger. So *Quetzalcoatlus* must surely go down in history as having the longest fourth finger ever to have evolved in any group of animals, this single digit probably reaching a staggering five meters in length. The wing-supporting fourth finger in the earliest known pterosaur, the Late Triassic *Preondactylus*, was relatively shorter than in later pterosaurs. Moreover, the relative sizes of the finger bones indicate that their elongation was greater in those elements farthest from the body.

Gleaning information on growth rates in extinct groups like pterosaurs is clearly a very difficult exercise. It is likely that the peramorphic acceleration in finger length occurred in a similar fashion to that in bats. In recent years the study of juvenile pterosaur growth has intensified, partly from increases in research, particularly by Christopher Bennett of the University of Kansas, and also on the basis of new discoveries of juvenile material, such as a newly discovered pterosaur "rookery" in Chile. In this deposit, found by Mike Bell of Cheltenham and Gloucester College in England, thousands of bones of young individuals were discovered high in the Chilean Andes in a two-meter-thick conglomerate. Investigation of the bone structure of these baby pterosaurs by Kevin Padian of the University of California revealed that bone formed early in development was packed with blood vessels. This indicates very rapid early growth. The absence of adult bones in this deposit suggests, first, that pterosaurs nested in large numbers, and second, that these young reptiles were incapable of flying. Presumably the jumble of rock and bones was produced by a flash flood of some sort. Adults would have been able to fly away. The juveniles could have escaped only if they were capable of flight. Their fossilized remains suggest that they weren't.

Bennett has observed a similar, highly vascularized bone structure in other pterosaurs, such as the Jurassic *Rhamphorhynchus* and the Late Cretaceous *Pteranodon*.[21] From the many specimens of *Rhamphorhynchus muensteri* that he examined, Bennett was able to show that the pterosaurs fell into three age classes, based not on size but on morphological features. Immature juveniles show a number of unfused skeletal elements and incompletely ossified bones. Older juveniles had some fused bones; others are more ossified. The largest specimens, which had an estimated wingspan up to 1.8 m, have all the bones ossified. Moreover, they are covered by a hard outer layer that lacked blood vessels, indicating that pterosaurs had determinate growth, probably stopping at sexual maturity. Bennett considers that the three size classes that he identified represent year-classes, and that each fossil sample represents a mass mortality. From this he has interpreted the age at which sexual maturity is reached, and growth stops, as three years. By identifying features other than size that can be used to indicate ages of animals when they died, it will be possible to begin to unravel the heterochronic processes that cause morphological differences between species.

Analysis by Bennett of wing finger length versus humerus length in *Rhamphorhynchus* shows that the wing finger is less than ten times the humerus length in juveniles, but roughly ten times humerus length in subadults. Seemingly, as in bats, this finger underwent its great elongation after birth. The extent to which it could grow in the egg would have been very limited. It is probable that newly hatched pterosaurs were incapable of flight. With the likely high levels of predation encountered by young pterosaurs, the faster this wing finger grew, the faster could it take to the air.

The evolution of many a structure, be it a wing or a petal, often carries with it a direct adaptive advantage. Maybe there was something about that shape, or that size, or that combination of shape and size, that in some way or other conferred a survival advantage on the individual or group of individuals—that made them just that bit more effective in surviving in life's lottery. But any evolutionary change in shape and size is more often than not likely to result in a change in the organism's behavior. A Paleocene nocturnal glider may have been able to do just that: glide from tree to tree. But an Eocene bat that had developed the power of flight was altogether in another ballpark. Similarly, tinkering with the time of onset of sexual maturity will engender a smaller or a larger body size, depending on the direction of tinkering. It too is liable to have a behavioral impact: smaller species can hide better than larger species, but larger species may be more effective killers. Selection for particular morphologies may sometimes have as much to do with a change in behavior, or with a response to changes in life history strategies, as it does with changes in function. Tying together the complex, interwoven pattern of the consequence of changing ontogenies, in terms of size, shape, behavior, and life history strategy, will be pursued in the next chapter.

10

Evolving a Way of Life

"A dog's a wolf that hasn't fully developed. It's more like the foetus of a
wolf than an adult wolf, isn't that so?"

Pete nodded.

"In other words," Mr. Propter went on,
"it's a mild, tractable animal because it's
never grown up into savagery. Isn't that
supposed to be one of the mechanisms
of evolutionary development?"

Aldous Huxley, *After Many a Summer Dies
the Swan*

IN NORTH AMERICAN DESERTS THERE CAN BE FOUND THE KANGAROO
mouse and the kangaroo rat. Not marsupials, but geomyoid rodents.
Both have a relatively large head, huge hind feet, large eyes, and a
long tail. Such anatomical features were long perceived as being
specific adaptations to living in a desert. The large head was seen as
counterbalancing the rodent while it hopped purposefully through
the cool night air on its huge, sand-paddle feet, searching far and
wide for food with its big eyes, and steering with a long, rudder-
like tail. The quintessential adaptationist explanation, of course, is, to
each structure, a purpose; for each purpose, a structure. But animals
are more than just constructions of biological Lego. How they evolve
has as much to do with how they live, how they reproduce, how many
offspring they produce, their metabolic rates, and their behavior as it

does with the structures from which they are constructed. Do the kangaroo rats and mice use each part of their anatomies for a specific purpose? Or, as some have argued, are some of these features little more than evolutionary baggage—anatomies with which they are burdened, maybe; but which have been of some use, somewhere, sometime? What of their lifestyles—their life spans, their methods of reproduction, their rates of growth—all aspects that can evolve and have profound effects on these rodents? Can they not also be the targets of selection, as much as the size of their feet and the color of their fur? If so, how can these components of their life histories and their behaviors evolve? Can the endless shuffling of ontogenies have any effect?

Protist Lifestyles

One of the major components of the marine biota is the plankton—a microscopic world of tiny monsters that if a thousand times larger would be the very stuff of nightmares. Many of these are the free-swimming larvae of marine invertebrates, such as echinoderms, corals, bivalves, and brachiopods (zooplankton), and bear little resemblance to their parents. But others are full-grown adults in their own right. Many are not animals at all, but protists—tiny, single-celled organisms, both animal-like (protozoans) and plant-like (chrysophytes and dinoflagellates). Among the protozoans, the best known are the foraminifers, probably because of their ability to produce an external shell, sometimes under their own steam, by secreting a calcium carbonate shell or by gluing sand grains around their bodies. Because of the possession of hard parts they are readily fossilized, and so have a rich fossil record. Occurring usually in vast numbers, they are a common element of many sedimentary rocks and play a particularly important role in biostratigraphy—they are free swimming, disperse widely, and in many lineages have speciated rapidly. A common feature of many foraminiferal lineages is a trend toward increased size. But what could the selective advantage of this be? Do we look to the advantage of the size itself, or should we be looking more at the conditions under which different sizes preferentially evolve?

These changes in size and complexity are under the control of heterochronic processes. Kuo-Yen Wei of Yale University has worked out the size and shape changes in an evolving plexus of species of the planktic foraminifer *Globorotalia*. Over the last three million years there has been a predominance of peramorphoclines, forming initially by predisplacement but later by acceleration as the foraminifers became more complex. Other lineages show peramorphosis by hypermorphosis, as the species got bigger, as well as increasing morphological complexity. However, at other times some lineages have gone the other way and followed paedomorphic trends. Wei ties these changes to fluctuating environmental conditions, in particular paleoceanographic events, which have modulated the morphological changes, rather than selecting for specific shapes or sizes that confer an advantage. The sort of heterochronic changes that occurred in such foraminiferal lineages, like heterochronic changes in any animals or plants, involve changes not only in growth rates but also in timing of onset of maturity.[1] Such factors play an important role in what have been termed "life history strategies." These include size at birth; growth rates; age of maturation; body size at maturity; number, size, and sex of offspring produced; and length of life. Attempts to categorize sets of traits have met with limited success. The most significant attempt produced what is known as the "*r-K* continuum." This is a descriptor both of environments and of the life history traits of the organism that lives in the environment.

An *r*-selected environment is unpredictable and often ephemeral. Here there is strong selection pressure for organisms to mature rapidly and so have short life spans (and so rapid cycling of generations), small body size (as a consequence of rapid maturation—in other words, classic progenesis), and large numbers of offspring. In environments such as ephemeral lakes that spring up in deserts after rain, short periods of plenty are interspersed with long, lean periods, during which mortality is high. In such environments selection favors progenetic paedomorphs that reproduce rapidly, producing lots of offspring. Short generation time enhances the chance of surviving long enough during the short, good times, when resources are in plentiful supply to allow reproduction of the next generation. Selection does not focus on particular anatomical traits but on the organism's ability to reproduce quickly. The side effect may well be a small body size and some peculiar anatomical features. But so long as they are not maladaptive for the organism they can be carried along as so much evolutionary baggage, provided that the baggage is not too heavy. If it is, the cost of excess baggage can then be enough to drag the organism to its doom, and to extinction.

In contrast, K-selected populations are those that inhabit constant, predictable environments. There is little random environmental fluctuation, so there is a crowded population of reasonably constant size. Competition among males is high. The characteristics of organisms in such environments include large body size; delayed onset of reproduction (by hypermorphosis); slow growth rates (by neoteny); and the production of few, large offspring. Many scientists have argued that the r-K continuum is far too simplistic. Certainly, it is often hard to make it work at the level of populations, although there are some instances where organisms have some r and some K characteristics. But at a higher taxonomic level, r-K patterns often seem to work. Of the paedomorphic processes, progenesis can be seen as being linked with many r-selected characteristics, and neoteny with K-selected. Of the peramorphic processes, hypermorphosis will be a principal mechanism enabling K characteristics to be selected. On the other hand, acceleration is likely to be found in r-selected environments. So, identifying an organism as having predominantly paedomorphic or peramorphic traits does not predispose it to any particular life history strategy. Heterochronic mechanisms are what count.

Viewed in this way, some of the heterochronic trends seen in groups like the microscopic foraminifers can be accommodated within such a life history model. Michèle Caron and Peter Homewood, working at the University of Fribourg in Switzerland, interpreted many of the patterns of morphological change that they saw in foraminifers in terms of r and K selection. During periods of environmental stress, paedomorphic species invade the oceanic surface waters, an r-selected environment. In other words, there are blooms of copious numbers of small, morphologically simple forms. Having short life cycles, and consequently simple (and what Caron and Homewood describe, interestingly, as "less evolved," by which they mean less complex) morphologies, these forms inhabit shallow waters, where the environment is less predictable. Deeper water species are K-selected, having longer life cycles, being relatively hypermorphic, larger, and morphologically more complex. Not only do characteristic shapes and sizes allow interpretations to be made of past water depths from fossil deposits, but the effect of changing sea levels will have constrained evolutionary trends, driven by changes in environmental stability and fueled by heterochrony. During periods of high sea level, the ocean depth on the shelf is greater, and so *K*-selected forms predominate. The seas are warmer; cold, deepwater upwellings more sluggish; and the environment on the whole more stable. A drop in sea level is often associated with reduced temperature, mixing of water masses, and upwelling of cold water from depth, all of which induce blooms of small numbers of opportunistic, progenetic species with short life cycles, adapted to a more stressful, unstable environment.[2]

The evolution of larger, benthic foraminifers has been similarly greatly influenced by life history traits. However, Jonathan Bryan of Okaloosa-Walton Community College in Niceville, Florida, has proposed that in these organisms another facet of life history, known as *stress selection*, has played an important role.[3] Stress selection occurs in environments that, while stable, are also stressful. Under such conditions the environment favors the selection of late maturers (by the operation of hypermorphosis) and very slow growers (by neoteny). These two processes often occur together, to such an extent that many definitions of neoteny have included the fact that often neotenic individuals are larger than their ancestors. Strictly speaking, they shouldn't be. The large size arises because hypermorphosis is occurring concurrently with the neoteny. In other words, not only is there a slowdown in the growth rate, but there is also a delay in the onset of maturity (see Chap. 12 for its impact on human evolution). An example of these processes operating in a stressful environment is the giant panda, its slow growth arising, it has been suggested, from the poor nutritional value of its bamboo food source (the stress), although the provision of the food source has (until recent times and the intervention of humans) been stable.[4]

In the case of large benthic foraminifers, trends along an environmental gradient from shallow, high energy/high illumination (the foraminifers living in a symbiotic relationship with an alga) to deep, low energy/low illumination show differences in growth rates and timing of reproduction. In shallow water there is rapid growth, but the environment is not ephemeral, so there is no selective pressure to reproduce early. Growth is prolonged and reproduction delayed, the larger body size arising from hypermorphosis. These forms are *K*-selected relative to deeper water forms. In those living in deep water the lower light levels reduce the efficiency of the algae, with the result that selection favors large juveniles that will inherit symbiont-rich protoplasm from their parents. Growth is also prolonged by hypermorphosis, but the growth rate is reduced (neoteny), because the low light levels and hence reduced nutrient levels induce stress selection. So, here we can see how the intimate intertwining of environmental factors and intrinsic growth rates and durations mold the species.

A Forty-Million-Year Journey

The first attempt to correlate heterochronic changes with environmental gradients in terms of life history strategies was undertaken by Mike McKinney of the University of Tennessee at Knoxville, on the basis of an analysis of the ontogenetic development in a number of lineages of Eocene sea urchins from Florida. What McKinney found was that fifteen of seventeen lineages showed trends of increasing size as they evolved into deeper water. This had occurred either by slower neotenic growth or by hypermorphosis. This implied that selection was targeting factors that McKinney related to evolution into more stable, deeper water, *K*-selected environments. McKinney considers that the size increases and attendant shape changes were brought about by a mosaic of heterochronic processes, the shape changes being, to a large degree, incidental by-products of selection for a particular ecological strategy.[5] Having, myself, studied evolutionary trends in Tertiary urchins from Australia, I attempted to reproduce Mike McKinney's findings with my material, since I had identified a number of trends in different urchin lineages. For some years I was mystified by the fact that the pattern of trends toward increased size and hypermorphic or neotenic traits in deeper water lineages, so clearly demonstrated in the Florida echinoids, saw no parallel in the Australian material. There seemed to me no reason why the ecological patterns operating on one side of the world shouldn't also have been occurring on the other side. However, there is a basic difference in the environments in these two regions. One, in North America, had remained at roughly the same latitude throughout the Tertiary. However, the other, Australia, had experienced a great journey. The key to understanding what was going on takes us on a journey on a rafting continent over tens of millions of years and brings in another aspect of selection for particular life history traits. This one involves ecological strategies adopted by juveniles, which are thus determined by heterochronic changes occurring not at the tail end of development but at the very beginning.

The clue to the very different patterns of selection of particular life history strategies in the same group of animals on both sides of the world may well lie in the tiny marsupiate (direct-brooding) urchins that I found in 40-million-year-old Eocene rocks in southwestern Australia (see Chap. 7). Sea urchins, like other echinoderms such as starfish, display a wide range of juvenile lifestyles, even though the adults may resemble one another very closely. A typical (if there is such a thing) urchin after fertilization initially forms an embryo, then develops into a pluteus larva. This is a free-living stage that looks nothing at all like a sea urchin and that swims in the plankton and feeds on unicellular algae. The larva descends to the substrate, undergoes a metamorphosis, and turns into a minute replica of the adult urchin, which feeds in a completely different manner from the larva. It then just grows larger, producing more spines and tube feet. Such a life history strategy is thought to be the "primitive" type.

Many urchins, however, miss the planktic larval stage altogether and produce either a nonfeeding, free-living larva or a larva that is brooded by the adult. These tiny replicas of the adult either nestle between the protective spines of the adult or sit in specially formed pouches in the urchin's shell, known as marsuparia. The eggs produced by urchins with a nonfeeding (or lecithotrophic) mode of development are much larger (350–2,000 μm) than eggs of urchins with feeding, free-swimming, planktotrophic larvae (70–250 μm). According to Greg Wray of the State University of New York at Stony Brook, the large eggs arise from peramorphic egg production, developing either for longer or at a faster rate.[6] Large eggs are necessary to provide nutrition to the nonfeeding larvae. Such larvae take a much shorter time to metamorphose than the planktotrophic larvae. What happens is that some larval structures are either deleted or very much reduced. Mouths and guts aren't needed, so they don't develop. Adult structures also appear relatively earlier in lecithotrophic larvae. So we have the operation of sequential progenesis, resulting in earlier initiation of adult characters and the loss of certain larval features.

Wray argues that these changes in early development play a pivotal role in determining the evolution of urchin life history strategies. One of the trade-offs of a shortened larval life and the production of larger eggs is reduced fecundity and greater maternal investment, particularly in the brooding species. One of the reasons given to explain the advantage to the urchins of missing out on the planktic larval phase is that it is during this period that mortality rates are particularly high. While direct developers may lose out in terms of less wide dispersal, they gain in having lower mortality rates.

Today the vast majority of direct-brooding urchins occur in Antarctic waters. For example, of twenty-eight known living species of marsupiate urchins, twenty-five live around Antarctica. It has long been argued that this type of juvenile development was a specific adaptation for inhabiting very cold water. This interpretation was used as a means of interpreting past oceanic temperatures by using the diversity of brooding urchins in the fossil record. Brooders that have marsuparia are readily identifiable in fossil material. Lots of marsupiates at any geological horizon was taken to mean colder water conditions when the sediments were deposited. Because the Tertiary rocks of southern Australia have the richest marsupiate urchin fauna in the world (twelve out of eighteen), I undertook an analysis of the changing diversity of these urchins through sedimentary rocks that had been deposited from 40 million years ago up to the present day. I then compared these changes with independently gleaned paleo-temperature data to see if there was any obvious correlation. What the analysis revealed was that there is no correlation at all between high marsupiate urchin diversity and low oceanic temperatures. But the diversity certainly changed appreciably. From the earliest species of marsupiate urchin that I found in the 40-million-year-old Late Eocene rocks in the quarry near Albany in Western Australia, the numbers of marsupiate species progressively increased over the following 20 million years, before undergoing a steady decline. Today only one marsupiate species is known to inhabit southern Australian waters. Yet this decline occurred as the Southern Ocean temperature dropped dramatically as the world plunged into the most recent Ice Age.[7]

The answer to what was controlling this change in diversity of marsupiate species, and which so affected the type of juvenile mode of development, is found in Australia's global wanderings. Paleo-geographically Australia has undergone a great journey over the last 40 million years. Forty million years ago Antarctica was moored off the southern coast of Australia. The southern Australian coastline to-day is at about latitude 35°S. Then it was at about 55°S. This means that 40 million years ago Australia was nearly 2500 km further south than it is today, deep in Antarctic waters, although at that time they were much less cold than they are today. Andrew Clarke of the British Antarctic Survey in Cambridge, England, has studied many of the invertebrates that occur in Antarctic waters today. Clarke has argued that direct-brooding invertebrates occur frequently in these high-latitude waters, but not because of the low temperatures; their presence is a reflection of the life histories necessary to cope with life at high latitudes. In addition to direct development, the larval life history strategy is characterized by slow developmental rates, long life expectancy, and low numbers of offspring. The combination of neotenic and hypermorphic characteristics results in many Antarctic invertebrates being large. Just take a look at the giant isopod *Glyptonotus antarcticus*, that grows up to about 10 centimeters long. It is symp-tomatic of the general trend of increasing body size with increasing latitude, but it has nothing to do with adaptation to living in colder waters. More likely, stress selection is operating at high latitudes. This may also manifest itself in the operation of sequential progene-sis, resulting in the loss of certain larval stages and direct, brooded development.

The Antarctic marine environment is actually very stable and pre-dictable, particularly with regard to food supply. Ocean temperatures remain virtually constant all year (varying by only about 2°C annu-ally). The supply of food, although varying to the extreme, is highly predictable. Herein may lie the stress. During the summer months, when light levels are very high, with 24-hour daylight, there is a vast phytoplanktic bloom. This results in a zooplanktic population explo-sion and is the reason why whales migrate to Antarctic waters in the summer months. Copious amounts of nutrients are released into the water, and all animals that inhabit these waters tailor their lives to this short (eight- to ten-week) feast. For the rest of the year there will be next to nothing to eat—which is stable, but stressful. Thus, because of this seasonal variation in food supply, larval development is timed to coincide with these short bursts of nutrient production. The amplitude of this seasonal variation is greatest in the highest latitudes and is known to diminish toward lower latitudes. So as Australia drifted north at about 6 cm per year (that's at about the same speed that your hair grows) it would have moved into regions of less pronounced seasonality, and therefore less predictable food supplies. Food might be forthcoming throughout the year, but in a less controlled manner. Stress selection would have been less important. During the Pliocene, direct brooding echinoids declined in abundance and were replaced by nonbrooders with free-swimming, but nonfeeding, larvae. These dominate the southern coastal echinoid fauna of Australia today. Others have free-swimming, feeding larvae. Interestingly, one of the most studied pair of species comprises *Heliocidaris tuberculata*, which has planktotrophic larvae, and *Heliocidaris erythrogramma*, which is lecithotrophic. Although they arrive at metamorphosis by completely different larval routes, the two adults are very similar in appearance and occupy similar niches.

Returning to environmental gradients and particular life history strategies, it becomes clear that whereas the latitudinal position of the Florida region has not changed substantially over the last 40 million years, the great trek north undertaken by Australia has had a pronounced influence on the marine environment. I haven't even mentioned the effect that the initiation of the circum-Antarctic current that became established near the Oligocene/Miocene boundary had, about 20 million years ago. This had a tremendous impact on the Australian marine biota. Thus in Australia no clear, simple picture of tying urchin trends to environmental gradients and life history strategies can be drawn. The picture is much more complicated. Even so, the urchins and other marine invertebrates could only exist in certain environments by virtue of the heterochronic changes that occurred at various stages of development.

A Surfeit of Lampreys

The effect of alterations in the length of particular juvenile growth phases has also had a pronounced effect on the life history of another, somewhat obscure group of animals, whose main claim to fame seems to lie with their role in the death of a king. On 1 Decem-

ber 1135, King Henry I of England died in Lyons, France. The cause
of his death has been attributed to one of the more insignificant
members of the animal kingdom—lampreys. These jawless, boneless
chordates, 450-million-year-old hangovers from the dawn of vertebrate evolution, are said to have been consumed by the king with such
relish, and to such excess at one particularly extravagant banquet, that
they killed him. Still, this didn't quench the royal partiality for these
rather obscure fish. For centuries a Gloucester's Royal Pie was sent
annually to the king, replete with lashings of lampreys. One weighing
twenty pounds that was sent to Queen Victoria was decorated with
truffles and crayfish on gold skewers. On the top was a gold crown
and scepter, and at the base four golden lions and a banner with the
Gloucester coat of arms—some pie![8]

The Gloucester connection arises from its position at the head of
the estuary of the River Severn in England. Following about four and
half years feeding in the rivers as larvae (known as ammocoetes),
some lamprey species, such as *Lampetra fluviatilis*, metamorphose into
a parasitic adult form that migrates downstream and out to sea, where
they attach themselves to other fishes by means of a suctorial oral disc
and ingest the blood and tissue of their host. This is vastly different
from their juvenile life, which is spent as a blind, suspension-feeding
larva burrowed in the mud. In their seventh year they return to the
rivers to spawn, shortly after which they die. But not all lampreys pass
through this parasitic phase. Closely related pairs of species have
been identified (in the case of *L. fluviatilis*, it is paired with *L. planeri*),
one being parasitic, the other not.

The nonparasitic species have a very different lifestyle, because

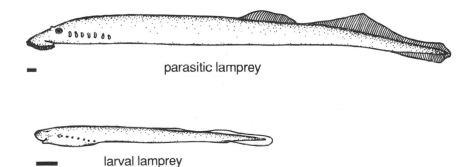

parasitic lamprey

larval lamprey

they delay their metamorphosis to the adult. When the juveniles of *L. fluviatilis* become free-swimming adults, the nonparasitic *L. planeri* continue living in the mud as larvae. They remain like this until the seventh year, when they finally get around to metamorphosing into the adult phase, ending up looking remarkably like the adult of their paired species. However, the nonparasitic species never develops the suctorial oral disc, so during its short (six to nine months) adult period it does not feed. In their brief adult life all they do is just spawn. Then they die.

The parasitic species of each pair is considered to be closest to the ancestral state, the nonparasites being interpreted as having evolved from them. For just how long some species of lampreys have had this propensity for acting like aquatic vampires is not clear. Lampreys have a long evolutionary history, extending back to their origins in the dim, distant Early Paleozoic, nearly 500 million years ago, at the dawn of vertebrate evolution, when all fish lacked jaws. The evolution of a nonparasitic paired species with such different life habits arises simply from a prolongation of the larval phase. This has the effect of pushing the parasitic phase out altogether, because time of spawning is the same in each member of the pair of species. Growth of the eggs progresses slowly in the parasitic phase, for it has about two years to bring them to viability. The eggs of nonparasites, on the other hand, have a much shorter period to become viable, so they undergo very rapid growth, ending up much the same size as the eggs of the parasitic species. There is little apparent morphological effect in the adults caused by this sequential hypermorphosis, this stretching out of the larval phase. Nevertheless, it has quite a profound effect on the life histories of the two species, confining one to life in the mud, the other to a more active phase as a mobile, fishy Dracula.

The Return of the Rodents

Opening this chapter I talked about the conundrum of trying to interpret the adaptive significance of every morphological trait in kangaroo rats and mice, and how this may have obscured the true nature of the targets of selection. To sort this out, it is necessary to compare the ecological strategies of closely related taxa. This way it might be possible to tease out the relative significance of anatomical traits as targets of selection, compared with life history strategies.

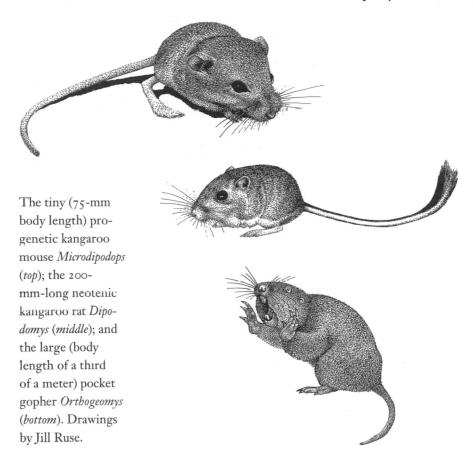

The tiny (75-mm body length) progenetic kangaroo mouse *Microdipodops* (*top*); the 200-mm-long neotenic kangaroo rat *Dipodomys* (*middle*); and the large (body length of a third of a meter) pocket gopher *Orthogeomys* (*bottom*). Drawings by Jill Ruse.

Mark Hafner, working at Louisiana State University, and his brother John, at Occidental College in Los Angeles, have attempted to do just this for the kangaroo mouse (real name *Microdipodops*) and the kangaroo rat, *Dipodomys*. Both of these geomyoid rodents possess a relatively large head, big hind feet, large eyes, and a long tail. But rather than these characteristics being specific adaptations to a desert life, the Hafner brothers have argued, selection was actually targeting the rodents' ecological strategies. Although they look alike, the lifestyles of these two rodents are very different.[9]

The kangaroo mice are progenetic and have all the characteristics that you would expect from *r*-selected animals, such as small body size, a short life span, and large litters; moreover, they live in an unstable, ephemeral desert environment. Kangaroo rats, however, have quite a different lifestyle, and their paedomorphic features are the product of slow, neotenic growth. In addition to slow development, they have a long life span, long gestation period, enlarged brains, and small litters, all features associated with K-selected animals occupying stable environments. The Hafners have argued that it is these life history strategies that are more likely to have been the principal targets of selection, rather than any specific morphological features. Obviously, they do not argue that the morphological features of the two rodents are of no adaptive significance—of course they are. Their point is that features of the rodents' anatomies are less likely to have been significant as targets of selection than as life history strategies.

Not all the features of kangaroo rats are paedomorphic, however. Enlarged, paired structures in the skull, known as the auditory bullae, and an elongated tail are relatively peramorphic, not having been subjected to the reduced rate of development that occurred in other traits. It is interesting that the kangaroo mice and kangaroo rats evolved their long tails in very different ways. The kangaroo rat has fewer tail vertebrae, but each is larger than in the kangaroo mouse. This implies there was acceleration in the rates of growth of individual vertebrae, producing a longer tail. However, the kangaroo mouse has more vertebrae, so it must have experienced an acceleration in the actual number of vertebrae that formed in the tail during early development.

Supporting evidence for lifestyles having been such important factors in the evolution of geomyoid rodents comes from other rodent genera. Not surprisingly, they are not all paedomorphic. The Hafners have highlighted the fact that some species are distinctly peramorphic, having extended their early rapid growth periods by delaying the onset of sexual maturity. The result—large rodents. The most striking of these is the pocket gopher, *Orthogeomys*—a veritable Arnold Schwarzenegger of the rodent world. This hypermorphic rodent is blessed with a heavily ossified skeleton, strongly fused skull, and large size—a tank of an animal among rodents. It has evolved "beyond" the normal shape and size attained by a generalized rodent. In terms of its lifestyle, the pocket gopher, like the kangaroo rat, is adapted to a K-strategic environment. Yet, this it attained in a quite different way, through extending its juvenile growth period by delaying the onset of sexual maturity, rather than by reducing its overall growth rate.

An obvious outcome of heterochronic differences between, say, the tiny, delicate, paedomorphic kangaroo mouse and the big, hefty peramorphic pocket gopher will be reflected not only in their life history strategies but also in fundamental differences in behavior. The effect of heterochrony on the evolution of behavior is perhaps one of the least appreciated areas of evolutionary theory, but nevertheless one of the most important.

Taming of the Dog

As the world was shaking off the grip of the last great advance of the continental ice sheets that had spread their shroud across much of North America, Asia, and Europe, a man living in what we now call northern Israel died and was buried. But this man was not buried alone, some 11–12,000 years ago. With him was interred a puppy. Whether it was the puppy of a dog or of a wolf, we are not sure. But it tells us that around the time that humans were beginning to forgo their hunter-gatherer existence and settle down to cultivating their own plants and tending their own animals, the links between a wolf-like ancestor of the modern domestic dog and humans were becoming established. Evidence that humans were beginning to associate with animals in a way other than to use them for their own materialistic needs, as food, or to cover themselves, or use their bones for tools, raises the question of how so-called domestication of animals like dogs actually evolved.

Darcy Morey of the University of Tennessee at Knoxville has posed the question of whether such early domestication of a wild animal was a rational decision.[10] Was it just an example of human subjugation of other organisms, as many prehistorians have argued? Did these early peoples make a conscious decision to selectively breed these animals, enhancing what were to them beneficial traits and eliminating undesirable ones—in other words, undertaking artificial selection? Or, as Morey has argued, should we look at the evolution of the domestic dog (for that is exactly what it is—evolution of what we now describe as a distinct species, *Canis familiaris*, in just a little over 10,000 years) as just another example of evolution by natural, rather than artificial, selection? If this is the case, then how did such a rapid transformation, from a large, ferocious wolf to (in most cases) a benign, simpering pooch, occur, in geological terms, so rapidly? And what influence did the selective pressures on particular morphological features have on the behavior of the domestic dog, compared with species of wolves?

If not a conscious decision, which intuitively seems unlikely—"Hey Gog, let's take that very vicious wolf that currently has its fangs plunged deep into my thigh and breed him with another equally unpleasant wolf and see if we can't just get a nice cute puppy that will sit on my lap rather than use me as its evening meal"—what was the setting under which some particular wolves gained some sort of selective advantage from associating with humans? Skeletal remains of dogs from archeological sites date back to about 14,000 years ago. The people living at this time were still hunter-gatherers. There is little argument that dogs evolved from the wolf, *Canis lupus*. Studies of DNA sequences in modern dogs indicate that genetically there is little difference between dogs and wolves. The two can mate and produce fertile offspring. (This begs the question of whether the dog, *Canis familiaris*, should really be considered to be a species distinct from *Canis lupus*—are we really harboring wolves in our homes?) The profound differences in shape, size, and behavior are a consequence of simple perturbations in developmental timing and growth rates.

Since these late Ice Age hunter-gatherers and wolves would have both been hunting the same prey, they would have come into frequent contact with each other. Wolves, especially wolf pups, will also opportunistically scavenge, so their contact with people would have been further strongly cemented. Morey envisages a scenario whereby wolf pups would have essentially been adopted into the human group. Studies of socialization in dogs show that this is best achieved early in the dog's life, between three and twelve weeks after birth.[11] These first few weeks are critical for establishing primary social bonds. Pups more amenable to human contact (and to human dominance) are more likely to prosper. Those wolf pups that didn't bond with humans, or who displayed more aggressive behavior, would have been either killed or driven away. Unwittingly, juvenile behavioral traits would have been selected by the hunter-gatherers.

At some stage these "domesticated" wolves would have begun

to change physically and behaviorally. Morey's examination of dog skeletons from archeological sites shows that dogs became smaller; in particular, their snouts shortened, producing a shorter-faced canid with steeply rising forehead and relatively wider head. Thus, we see the same changes as I outlined in Chapter 4, the selection, in more recent times, of smaller, more paedomorphic breeds of dog. But whereas artificial selection over the last few hundred years has been for either paedomorphic or peramorphic traits in different dog breeds, early in dog evolution, natural selection favored paedomorphic traits.

There are many arguments as to what were the factors that favored the selection of these more juvenile traits. Maybe it was the more "cute" juvenile appearance; or maybe the smaller size. Maybe some other features were unconsciously being selected. Morey favors more general aspects of the animal's life history strategy: reproductive timing and body size, producing a smaller, paedomorphic dog. Measurements of cranial characteristics of dogs from archeological sites show that they converged more and more on juvenile wolves.

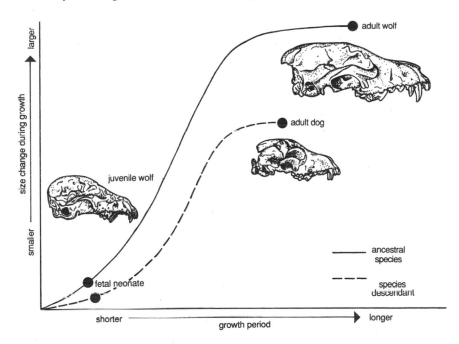

How the domestic dog has "evolved" from the wolf by paedomorphosis.

As Morey has stressed, compared with wolves, adult dogs not only look juvenile, they act that way too. They grovel, they simper, they whine, they play, and they bark a lot. Tameness is essentially submissive behavior, the hallmark of juveniles. Further evidence to support the contention that maybe juvenile behavioral traits were unwittingly being selected comes from experiments carried out in the 1960s by the Russian geneticist D. K. Belyaev on silver foxes bred in a commercial farm. Belyaev found that while most foxes were very fearful of people, about 10 percent were less so. By breeding those calmer individuals, over about 20 generations the product was foxes that actively sought human contact—licking people's hands, whining, and wagging their tails in a most doglike fashion. Significantly, there was a shift in reproductive cycles in the foxes, from one a year to more frequently.[12] Whereas wolves breed just once a year, dogs breed more often. What these experiments reinforce is how closely physiological and behavioral changes are intertwined. Moreover the two change in tandem, not by profound genetic upheavals, but by relatively simple alterations in the rates and timing of development. There is no need to invoke active artificial selection of dogs by Late Pleistocene hunter-gatherers. These people's own lifestyle, the niche that they had created for themselves, was a niche which dogs themselves could take advantage of—but only by evolving juvenile canid traits in the adults.

Changing Behavior

The evolution of dogs is just one example showing that often changing patterns of behavior are a direct consequence of evolutionary changes in morphology and life history strategies. Like morphology, an organism's behavior changes as it grows up, either by a series of discrete changes or gradually. Discrete behavioral ontogenetic changes accompany metamorphosis in amphibians, when they change from an aquatic larval phase to a terrestrial phase. Not only do major anatomical modifications occur, but the behavioral changes are profound. These distinct behaviors occurring at different stages of development are a direct consequence not only of the anatomical changes •but also of neural differences, as the brain size increases. Thus, the repertoire of behaviors available to a tadpole will be appreciably less than those that can potentially be undertaken by an adult frog or salamander. But paedomorphic retention of larval anatomical and environmental characteristics, as in the axolotl, is also accompanied by some paedomorphic behavioral traits. However, this is not an inviolate rule. Living salamanders, for instance, which show much evidence of paedomorphic morphological reduction, do not engage in correspondingly simplified behaviors. While the loss of lungs and larynx in many salamanders means that, unlike frogs, they do not have a complex system of sound communication, they make up for it by employing pheromone communication in a much more sophisticated way.

So, like morphology, behavior can also be affected by changes in the timing and rate of its ontogenetic development. If a species' ontogeny is prolonged, compared with its ancestor, there are likely to be associated behavioral consequences. If ontogeny is foreshortened, then ancestral juvenile behavioral patterns can be retained by descendant adults. Similarly, the rate of change of behaviors during ontogeny can be sped up or slowed down, or the transition from one behavioral stage to another can be delayed or brought forward, just like morphology. So the three groups of heterochronic processes apply equally well to the evolution of behavior as they do to the evolution of morphology. Such heterochronic changes have the capacity to have a major impact on the species and, it can be argued, even be the principal targets of selection.

A species' behavioral repertoire is a consequence of the extent of its neural development, combined with anatomical complexity. Because, in general terms, greater neural and anatomical complexity go hand in hand, so too does behavioral complexity. Behavior comprises a mixture of "lower-level" or genetically programmed responses (such as reflexive and instinctive behavior) and "higher-level" behaviors, such as learning and reasoning. Lower-level behavioral changes are very closely tied to genetic changes that influence the timing and rate of growth. In ants, for example, there is a direct correlation between behavioral complexity and head size (reflecting brain size). The larger and more complex the brain, the more complex these behaviors.[13] The same is true of mammals.

One part of the ant brain, known as the *corpora pedunculata*, plays a major role in selecting motor programs and the formation of aspects of behavior, such as learning. This region increases in size as the overall size of the brain increases during growth. So, in those species that have evolved a larger head, and consequently have a larger brain, the larger corpora pedunculata provides the capacity for more intricate behavioral patterns. Any evolutionary processes that promote either an increase in the rate of growth of the head or a prolongation of the period of its growth will result in the evolution of species capable of performing more complex behaviors. It is known, for instance, that small ant species perform fewer categories of behavior than larger ants. Because larger ants have larger brains, they also have a greater number of neurons, allowing a wider range of behaviors to be expressed. The question is, has the evolution of ants with larger brains and larger body sizes occurred because, being able to undertake a greater range of behaviors, they are somehow superior to smaller ants? Maybe. But selection for large body size in ants is also influenced by a host of other factors, such as nest size, potential predators or competitors, food type, and colony size. But whatever factor influences the evolution of a particular body size, brain size, and thus the extent of the behavioral repertoire, is firmly locked into the equation. However, given that head size could, up to a point, increase more than the rest of the body, it is likely that selection acts on a suite of characteristics, only one of which is behavior. Such behavioral differences are not only to be found between species. Polymorphism (the formation of different types of individuals within a species, such as soldier and worker ants) often arises as a consequence of variations in developmental times. For example, a delay in a larva metamorphosing into an adult results in individuals developing as larger soldier ants, rather than smaller worker ants that metamorphose earlier. With the larger body size comes a bigger brain and more sophisticated behaviors.

More complex behavioral patterns, those that involve learning and reasoning, can also be affected by heterochrony. This is particularly so when extensions or contractions of the organism's ontogeny result in changes in adult brain size. For this will cause changes in the number of neurons that store and process information. Variations in duration of the juvenile growth period also mean variations in the length of the learning period. Thus, any hypermorphic delay in the onset of sexual maturity results in a longer childhood and often, in "higher" mammals such as primates, a longer period of parental care during which the juvenile can learn more complicated behaviors from the adult. But the advantage lies not only in a longer period for learning, but also in the development of more neurons for making most efficient use of this longer learning phase. This has been of particular importance in human evolution (see Chap. 12).

Ontogenetic behavioral changes have not been particularly well studied in many groups of organisms. However, one exception is birds. An early, classic study was undertaken by Herbert Friedmann and described in his book, *The Cowbirds*, published in 1929. Friedmann showed how the courtship displays in these parasitic birds change as the birds grow toward maturity. In some species the courtship behavior is essentially like that occurring at an earlier ontogenetic stage in a closely related species. The bay-winged cowbird from central South America has no courtship display at all. Both the screaming cowbird, from the same region, and the shiny cowbird, which is found over much of South America, have some rudimentary courtship display, fluffing out their feathers and bowing forward with arched wings. However, the North American cowbird does all these things and more—when it bows, it transcribes an arc of nearly 180 degrees. What Friedmann found, when he kept two young North American cowbirds in captivity, was that as they approached maturity, they passed through each of the stages attained by the South American species.

The same goes for the cowbirds' song. In the bay-winged cowbird the song is structureless, whereas the adult screaming and shiny cowbirds develop a more complex song pattern, although less so than the North American cowbird. Even in terms of coloration, the bay-winged has more primitive plumage than the other species. Even the extent of parasitism varies between the species. If the least parasitic is regarded as being closer to the ancestral condition, then again, the sequence of bay-winged to screaming, shiny, and then North American reflects an increase in specialization and complexity brought on by peramorphosis. While the bay-winged uses other birds' nests for its eggs, it does incubate and rear its own young. It will even, on occasion, make its own nest. The shiny cowbird has relatively poorly displayed parasitic behavior; it too will also try to make a nest occasionally. The North American cowbird, though, is fully parasitic, not only laying eggs in other birds' nests, but also leaving the nest builder to rear its young for it.[14]

There has been much conjecture over the years as to how such parasitism arose in the first place. But as long ago as the the first decade of this century, one Professor F. H. Herrick provided an interpretation based on what is essentially heterochrony of behaviors. He proposed that parasitism in cuckoos in Europe, and in the cowbirds in America, arose from relative changes in the timing of expression of egg laying and nest building. He found that in the black-billed cuckoo, which is not parasitic, the timing of these two behaviors is coincident. In the parasitic European cuckoo, however, they got out of synchrony. As Herrick wrote, "The door is thus opened wide to parasitism in its initial stage, whenever the acceleration of egg-laying or the retardation of the building instinct becomes common, with or without irregularity in the egg-laying intervals."[15]

Other nesting behaviors can also arise from changes in the normal sequence of ontogenetic events. The nesting behavior in birds that nest communally can have the effect of prematurely inducing feeding behavior.[16] What happens is that the juveniles or nonbreeding young adults can have their feeding, caring, and nest-building responses turned on at an earlier developmental time due to the presence of nestlings in the nest. For example, in pied kingfishers, juveniles or young adults who are sons of at least one of the breeding pair will help in feeding younger offspring. In these communally nesting birds the helpers assist in feeding, nest building, and resolving territorial disputes. These are normally adult behaviors, but helpers will be either nonbreeding adults, subadults, or juveniles at different stages of development.

Some of the best-known cases have been described in Australian fairy-wrens. The males of these birds undergo an ontogenetic change in plumage from brown as juveniles to developing a blue tail toward the end of their first year and then full blue plumage. In their second year, this blue plumage is lost for nearly half a year when they are not breeding. But in the third and fourth years the blue plumage stays for longer, becoming permanent in the fourth year. In the communal nests, usually only one male has this full blue color. Many of the brown birds with them are not the females, who never develop the blue plumage, but young males. They remain as helpers, undertaking breeding adult behaviors, such as feeding and caring for the nestlings. The breeding female will be the only one, however, to brood the eggs. In fact, the nonbreeding males and the breeding male stay to care for the young while the females leave to renest elsewhere after the eggs have hatched. Such communal behavior enables rapid population growth, by allowing repeated nestings when conditions are suitable.[17]

Australian choughs display a similar "helping" behavior. The white-winged chough live in groups of four to eight. Immatures tend to remain in the nest and help with nest building, even though they tend not to do it very well. These young birds will also help to feed both nestlings and other, younger fledglings.[18] An even more extreme example of precocious onset of adult behavior occurs in the Australian white-rumped swiftlet. Although more than one egg is laid in the nest, the parent will only incubate a single egg. After it has hatched, the nestling immediately adopts the parents' behavior and proceeds to incubate its sibling. Such behavior is thought to be under hormonal control and has probably evolved because the swiftlets' nestling period is very long, necessitating a long period of care, obviously more than the parents can provide.[19]

The genetic underpinning of such behavioral changes is illustrated by one of the more unexpected outcomes of the mating of wild and domestic geese, arising from the fact that the two types of geese take different times to reach sexual maturity, domestic geese maturing earlier than wild geese. A rather unfortunate consequence for male offspring from a mating between these two types is that they attempt to mate with their mothers. This confusion, producing an "Oedipus" behavior, is due to the earlier onset of maturity inherited from the domestic parent, but this is combined with a delay in the offset of the mother-following behavior, inherited from the wild parent. Consequently, the confined male goose wants to follow its mother dutifully, but is also compelled, due to its earlier onset of maturation, to mate with the closest female—unfortunately, in this case, his mother.

One of the more intriguing examples of the effect of heterochrony on behavior is the evolution of birdsong. Rebecca Irwin of the University of Michigan looked at how birdsong changes, both as individual birds grow up and between species. As young birds develop, their song changes through a set sequence of "arias." In Irwin's study of three species of sparrows, she found that the nestlings have a continuous song, which, as they grow to adulthood, changes to discrete songs. The repertoire is ill defined and the song syllables have little structure in juveniles, but are more structured into numerous songs as the birds grow. However, the extent to which these ontogenetic changes occur varies between the three species. Thus, for example, in the adults of one species, *Zonotrichia leucophrys*, the many songs sung by the juvenile resolve to a single song. In the other two species, both members of the genus *Melospiza*, they continue the juvenile behavior of singing many songs.[20]

Another paedomorphic feature retained in some birds is mimicry, usually a juvenile characteristic. These birds never attain the "crystallized" song of other adult species. They are forever learning new songs and forgetting old ones, not retaining any one particular song. The song slowly develops through the bird's life, forever changing until it dies. In reed warblers, however, song crystallization occurs at an earlier age than in related species, no new ones being learned after that. Retention of an otherwise juvenile song can sometimes have obvious adaptive advantages. Ron Johnstone, ornithologist at the Western Australian Museum, has recounted to me how the Australian cuckoo has a juvenile call that is the trigger for the adults of whatever bird species it is nesting in to feed it. Even though the cuckoo's song changes at maturity, it can still produce the juvenile call when hungry. During migration, adult cuckoos have been observed using the juvenile call, triggering a feeding response in other birds to provide it with a free meal!

The more we look at heterochronic changes in morphology and interpret them in terms of the organism's functional morphology, it becomes clear that the final outcome of these changes in shape, size, and life history is a change in behavior. The tiny progenetic trilobites that swam in the seas half a billion years ago evolved this ability to swim freely by retaining their ancestors' juvenile habit and losing their adult habit of crawling along the sea floor. An evolutionary change in size, as well as a change in shape, brings with it a change in behavior. Evolutionary tinkering with neural development can change a bird's song or an ant's aggressive behavior.

Let me finish with one of the more bizarre human behaviors that may well have its origins in the retention in a few adults of what is usually a juvenile condition. Two Russian composers, Rimsky-Korsakov and Scriabin, and one French one, Messiaen, can be counted among a very small proportion of people with a condition known as synesthesia. For these people there is a confusion of the senses: a sound can have a color or a taste; a color can also have a taste; or an image can provoke a sound. The degree of this confusion can be great. While my daughter, when playing a reasonably dissonant chord in a piece of piano music by Béla Bartók suddenly said, "Yuk, that tastes bitter!" other people have much more dramatic experiences. Daphne and Charles Maurer, in their book *The World of the Newborn*, report on studies carried out on one particular individual for over thirty years. This person saw high-pitched oscillations looking like fireworks; strips of color had a rough feeling and tasted of briny pickle; and the number 8 was seen as being milky-blue, like lime.[21] Experiments on three- to four-week-old babies show that this confusion of senses may well be "normal" in newborns. The babies were found to equate brighter lights with louder sounds. The usual line of neural development, which sees this confusion rapidly lost, may be retained paedomorphically by a small number of individuals, the particular early juvenile neural pathways persisting deep into adulthood. Who knows what other behavioral differences in humans and in other species owe their origins to developmental changes?

11

Fueling the Biological Arms Race

It must have been that in the early days of earth

Countless kinds of living things died out

And failed to reproduce their kind.

For all the living things you now see feeding on the breath of life

Must have survived, after the first appearance of their kind,

Either through cunning or through valour or through speed of foot.

From Lucretius, *De rerum natura*, Book V (c. 55 B.C.),
translated by A. D. Winspear

MENTION SEA URCHINS TO ANYBODY AND THEY IMMEDIATELY CONJURE up an image of a mobile, marine pincushion, but with the pins pointing out, rather than in—organisms that to all appearances seem almost as impregnable as Fort Knox. You could be forgiven for thinking that the sea urchin, possessing such a fearsome array of spines, some of which may be poisonous, is as safe from predation as any animal alive. Yet it was not for nothing that such defensive spines, as well as a variety of other antipredation strategies, evolved in sea urchins during the last 100 million years. To put it simply, sea urchins are mighty good to eat. As well as the predilection of many humans for a spoonful of urchin gonad, these poor creatures are known to be consumed by a wide variety of predators, including other sea urchins, starfish, gastropods, crustaceans, fishes, turtles, birds, sea otters, and even arctic foxes. The evolution of various antipredation strategies, which may involve improved armor or the ability to hide away from predators, but still allow the animal to feed and function effectively, is fueled by various heterochronic processes. But the driving force behind the directions that evolutionary trends take in urchins, and many other group of organisms, is predation pressure.

Setting the Trend

The long-playing fossil record sings us songs of innumerable evolutionary trends—of directional evolution, either in size or shape, often persisting for many millions of years. Sometimes these trends occur within a single species as it changes gradually over hundreds of thousands of years. In others, a number of species show a directional trend in some trait or other, marching resolutely in a single direction. This may occur in a jerky, stepwise fashion, as one species persists, essentially unchanged for a few million years, to be replaced in relatively rapid transition by another, this pattern repeating itself through a number of species, but in a certain morphological direction. But neither the species nor the lineage of species can have any sort of built-in purpose. Rather than evolving in a haphazard manner, they evolve in certain directions at certain times. This is because evolution is often tightly constrained—it is certainly anything *but* random—by the nature of the ontogenetic trajectories which provide one of the directional components. And it is constrained by another directional component, the environmental gradient along which the particular evolutionary trends track. But what gets these trends going in a particular direction, along a particular gradient, in the first place?

I have to admit that evolutionary trends in the fossil record have intrigued me for many years. As you know, I have become particularly interested in one group of animals, whose fossil remains occur frequently in sedimentary rocks in many parts of the world—sea urchins. In some quarters, my interest in such objects could be considered to be a trifle obscure. But fossil sea urchins have fascinated not just me and a few other paleontologists, but also many other people for perhaps longer than any other object that we know about. While to me their fascination lies in what they can tell us about the mysteries of evolution, to humans, as far back as those living about 70,000 years ago, who carved tools from fossil urchins, they must have been equally fascinating. Many archeological sites, especially in northern Europe, have yielded fossil urchins. In some instances they have been drilled through, probably to be used in necklaces, but in many instances irregular urchins, in particular heart urchins, have been found in burial deposits.

Perhaps the most poignant was the discovery, near Dunstable in England in 1887 by archeologist Worthington G. Smith, of the remains of an early Bronze Age female and child. Surrounding them had been placed more than two hundred fossil urchins. In other burial sites, such specimens have been found placed in strategic parts of the body, implying that great ritualistic significance was attached to them. Their presence in cremation urns in Iron Age sites in southern England, along with axe heads, forms part of a Norse myth: that both urchins and axe heads were thunderstones cast to Earth by Thor as he vented his wrath during a thunderstorm. In Celtic mythology, they

Early fascination with fossil sea urchins. A grave excavated by Worthington G. Smith in 1887 at Dunstable in England of a Bronze Age woman and child in which were found hundreds of fossil sea urchins. Engraving by Worthington Smith from his book, *Man—the Primaeval Savage*, published in 1897.

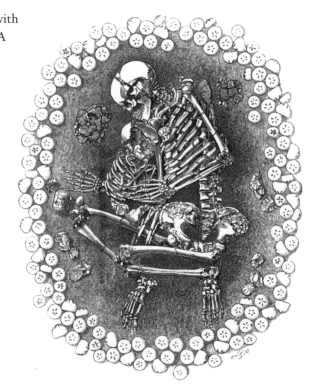

were considered to be snake's eggs and were objects of great magical power, being particularly prized up to the Middle Ages as powerful antidotes against poison. But the power of these objects today lies less in their purported medicinal properties and more in what they can reveal about the mechanisms of evolution.

Murder and Mayhem in the Miocene

Like most animals, the main concern of living sea urchins on a day-to-day basis is to stay alive. If they do this and also manage to reproduce, all well and good. This they achieve by eating and avoiding being eaten. Just look at the dazzling variety of shapes and ecological strategies that have evolved in this group of animals over the last hundred million years or so. While it is comparatively easy to observe the ravenous hordes of predators that attack urchins today, we must turn to the fossil record to see just how sea urchins have evolved different strategies to avoid being eaten and to keep one step ahead of their predators. Shuffling growth rates of the individual plates that make up the urchin's shell (known as the test), or of the number and size of the spines, has allowed various antipredation strategies to be adopted. Evidence of predation on fossils like urchins can be seen either from breakage caused by the teeth of fishes or the claws of crabs, or from holes cut by predatory gastropods. Selection on the prey under higher predation pressure will favor those forms that have evolved morphologies and/or behaviors that minimize the impact of predation. Prey can evolve smaller or bigger body sizes. Small body size allows the potential prey to hide more effectively and can also take it out of the predator's feeding range. Large body size can mean a thicker shell or more rapid locomotory behavior. A thicker shell can also take the prey out of the predator's feeding range, making it energetically inefficient for the predator to try to tackle a larger prey. Selection on the predator will tend to "follow" the prey—that is, if the selection favors larger prey, then the selection for larger predator will follow.

One particular urchin lineage that shows pronounced evolutionary trends fueled by heterochrony consists of three species of the heart urchin *Lovenia*, which lived in the seas off southern Australia between about 25 and 10 million years ago.[1] Unlike the typical spiky, reef-inhabiting urchin, such heart urchins, and other groups including sand dollars, generally have spines that are greatly reduced in size but greatly increased in number. These, plus other attributes such as a mouth situated near the front on the lower surface, and specialized tube feet that are used for respiration or feeding or mucus secretion, are all structures that have allowed these urchins to burrow into the sediment. *Lovenia* is a reasonably typical shallow-burrowing heart urchin. However, it has one character which is slightly unusual for heart urchins, but which occurs in a small group of genera, including *Lovenia*, and that is the possession of large defensive spines. These spines, similar to those that decorate the typical ancestral regular echinoid, are thought to have "re-evolved" sometime during the last 60 million years or so, arising from variations in their rate of growth during ontogeny. And it is the changing density and distribution of such spines in this *Lovenia* lineage, combined with varying predation pressure, that provides the key to understanding many of the changes that have affected sea urchin evolution.

The oldest species in this lineage, *Lovenia forbesi*, occurs in yellow limestone cliffs on the banks of the Murray River, near Mannum in South Australia. The tranquil beauty of this region belies the episodes of carnage that must have occurred when this area was under the sea 20 million years ago, for the rocks are studded with countless millions of sea urchins, many of which show direct evidence of their cause of death. Looking for all the world as though they have been clinically assassinated, the shells of many of the urchins are pierced by a single neat hole, which closely resembles that made by a small bullet. Yet by looking at the most likely culprit among some of the modern-day sea urchin's predators, it would seem that this assassin was far from being a swift killer. On the contrary, the urchin's death would have been a long, drawn-out process from which it had only a limited chance of escape. The killers were almost certainly members of the marine snail family Cassidae—helmet shells.

These snails today are known to feed almost exclusively on urchins, both nonburrowing and shallow burrowing types, and have a very distinctive way of killing their prey. When hunting for urchins, a cassid's siphon and tentacles protrude slightly beyond the shell. On approaching an urchin, both the siphon and the tentacles become fully extended. On making contact with its prospective meal, the cassid suddenly raises its foot and arches over its prey like Jack the Ripper before lowering itself over its victim. Having extended its proboscis, the cassid then commences its kiss of death. The end of the proboscis is armed with a radula that bears an array of cutting teeth. These are used to cut a neat disk out of the urchin's test. It is thought that the cassid may also secrete acid to aid this process. The deadly weapon is then inserted through the hole into the victim, and the slow death begins as it proceeds to suck out the urchin's gonads. Whereas the initial attack, following its period of stalking, may take just a few seconds, this feeding process may take up to an hour to accomplish. While there is little need to rush this activity, the preliminary rapid pinning of its prey is important because in cases where an urchin has been seen to be disturbed by a cassid and made off, it has always been the urchin that outruns the cassid. Cassids can attack both regular urchins, which do not burrow, and heart urchins, which are completely burrowed.

Whereas large (up to 35 centimeters long) living cassids have such a tough foot that the urchin's fearsome array of spines offers no defense, the spines that the fossil urchins from the Australian Miocene possessed actually helped protect them from their smaller attackers to some degree. This has been deduced by plotting the distribution of holes in *Lovenia forbesi*. Much of the upper surface of the urchin's test, where the spines occurred and which they covered, had fewer holes than did other areas of the test. A number of cassid species occur in these Australian Miocene rocks, and it is likely that these were the culprits responsible for the deaths. Data from living cassids shows that there is a close correlation between the size of the cassid shell and the hole that it cuts in its prey. The small cassids, such as 30-millimeter-long species of *Semicassis*, which coexisted with *Lovenia forbesi*, would have have cut holes about 2 millimeters in diameter. And this is the hole size found in the urchins. Analysis of many hundreds of specimens has revealed that almost 30 percent appear to have been killed by cassids. (Another heart urchin, *Eupatagus*, which coexisted with *Lovenia*, suffered almost 50 percent mortality from cassid predation).

Lovenia forbesi became extinct at the end of the Early Miocene and was replaced by another species. Whereas adults of *L. forbesi* developed defensive spines in eight of their ten interambulacral columns, they occur in only six columns in this Middle Miocene species. Juveniles of *L. forbesi* pass through a stage when they have spines in only six columns. However, where they do occur they are densely concentrated. This form also possessed more digging spines on its lower surface. Fewer specimens (only about 20 percent) were killed by cassids.

The youngest species in this lineage occurs in Late Miocene strata close to Melbourne and continues the trend seen in the other species: it has fewer defensive spines (occurring in only four columns, paedomorphically resembling early juveniles of the Early Miocene species); and a greater concentration of digging spines, reflecting a peramorphic increase in extent of development. Only about 8 percent of specimens of this species were killed by cassids. Moreover, healed cuts on the urchin's test occur far more frequently in this species, suggesting that the cassids attacking this youngest species were much less efficient at dispatching their prey than earlier cassids.

So why is there less evidence of predation on species with fewer defensive spines, if these were such an important part of the urchin's protective armor? A closer look at the sediments in which the urchins lived provides a clue. The oldest, most heavily predated species lived in a shallow water environment, for the sediment into which it burrowed was relatively coarse-grained. The younger species inhabited progressively finer-grained sediments, suggesting that they inhabited deeper, generally quieter water conditions than their ancestors.

By looking at the bathymetric distribution of cassids around Australia today, it can be shown that species diversity declines rapidly into deeper water, from 20 species between highwater mark and 100 meters, declining to five between 300 and 400 meters, to only one at depths in excess of 500 meters. The absence of telltale cassid predation holes on urchins from much deeper water Miocene deposits suggests that there was a similar paucity of cassids in deep water in the Miocene. So it can be argued that it was strong selection pressure in shallow water that selected initially for heavily spinose urchins. These evolved initially by localized peramorphic increase in growth rate of individual small spines. Subsequently, those forms which, by virtue of increased numbers of digging spines, were able to inhabit finer-grained sediments in deeper water were preferentially selected. In such an environment where there were fewer predators, there was less selection pressure to produce populations with many defensive spines.

As the sea urchin *Lovenia* evolved in southern Australia, there was a paedomorphic reduction in defensive spines. However, burrowing spines increased by peramorphosis as the three species in the lineage became better adapted to inhabiting finer sediments in deeper water. Reduction in number of gastropod-induced boreholes suggests evolution driven by predation pressure into deeper water, where there were fewer predators.

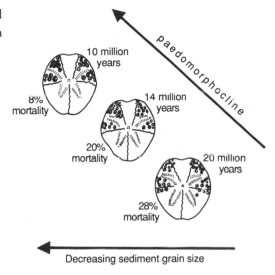

A similar pattern can be seen in living species of *Lovenia*. Those inhabiting deeper water have a sparser array of defensive spines than shallow-water inhabitants. Some of these living species show how another antipredation strategy has developed in this genus since the Miocene—they have grown larger. But so too have the living cassid species. This "arms race" struggle between coevolving predator and prey, with some species of *Lovenia* becoming too large to be eaten by cassids, has been matched by a corresponding selection for larger cassids. While Australian Miocene cassids were no larger than 70 millimeters (about twice the size of their prey), living species grow to almost 250 millimeters in length. Living species of *Lovenia* are three times longer than the largest of the Australian Miocene species, reaching up to 120 millimeters in length.

The onshore to offshore trend seen in the Australian Miocene *Lovenia* lineage is not an isolated instance. Many other lineages of heart urchins show a similar trend, perhaps also predation driven, but all fueled by heterochrony. Like *Lovenia*, many of these other lineages show a similar pattern of one species replacing another in a stepwise fashion—so-called anagenetic speciation. With such evolution there is no overall increase in species diversity, but a shift in overall morphological expression in the lineage and frequently evolution from one environment into another, in this case into deeper water. Since the Cretaceous, this onshore-offshore trend has occurred at higher taxonomic levels in a number of major groups of sea urchins, not just heart urchins. Analysis of onshore to offshore patterns in thirteen orders of post-Paleozoic urchins by Dave Jablonski of the University of Chicago and Dave Bottjer of the University of Southern California has revealed that seven of these show a trend of migration from onshore to offshore environments. Migration offshore may have occurred in two ways: by an expansion of some forms into deeper water, but with some representatives remaining in shallow water, or by a displacement, whereby there was a loss of onshore representatives.[2] Although the Australian lineage is an example of displacement, the genus *Lovenia* as a whole represents a case of expansion, for onshore representatives still exist today.

One of the classic examples of evolution in the fossil record is evolution within the heart urchin *Micraster*. The evolutionary changes that occurred in this urchin, which lived in the chalk seas of northern Europe 80–90 million years ago, have often been explained in the past as adaptations to burrowing deeper in the sediment. During the last 60 million years a number of heart urchins also developed morphological characteristics, facilitated by peramorphic changes, which allowed them to inhabit deep burrows. Some, such as *Schizaster* and *Brissopsis*, used this facility to inhabit muddy sediment. One particular adaptation was the evolution of mucus-secreting tube feet. These are able to line the funnel that connects the urchin to the surface. Others use this mucus to bind food they have obtained from the nutrient-rich surface layers with long tube feet, into strands like spaghetti. These are pulled down into the burrow and eaten. Predation pressure may also have been the selection pressure that so favored the evolution of such bizarre structures in inimical environments.

Earlier in sea urchin evolution during the Jurassic and Cretaceous, predation pressure, principally from fish and crustaceans on the normal, regular spiny urchins, was the main driving force, I believe, behind the evolution of burrowing irregular urchins. Furthermore, during the Cretaceous, about 100 million years ago, predatory marine snails evolved. Burrows are safe hiding places for urchins, away from the clutches of predators. However, in the Eocene one group of snails, the cassids, evolved. With their arrival the burrowing urchins were no longer safe in their snug little burrows, for these snails also had the ability to burrow. This resulted in selection pressure favoring those forms that were capable of deep burrowing, out of the reach of the cassid's searching tentacles, or that were able to live in deeper water where the cassids were unable to live (probably due to their inability to burrow effectively in muddy sediments).

And what of that most successful of urchins, the sand dollars (clypeasteroids)? These were the last of the major groups of urchins to evolve, the earliest appearing little more than 50 million years ago. This group has undergone a major evolutionary radiation during the Cenozoic. Do we owe the appearance of such groups merely to chance and the crossing of a major adaptive threshold into a major new niche? Or can the evolution of a major new body form (in this case a very flattened shape) be attributed to the minimization of predation pressure?

The earliest sand dollars were tiny paedomorphic forms known as fibulariids. These still live today, lurking between sand grains. The evolution of such small urchins was probably also in response to predation pressure, because another way to escape predation, apart from growing much bigger, is to be very small, either by maturing very early or by a strong reduction in growth rate. Not only does this help in hiding from potential predators, but for predators like cassids that can only feed on one individual at a time, there is no point bothering with such small fry. It was from such small fibulariids that all later sand dollars are thought to have evolved. There have been a number of suggestions as to why this flat shape should have been so successful. One is that it allowed the urchin to feed in the nutrient-rich layer close to the sediment-water interface, yet remain burrowed. The Australian Miocene rocks provide evidence that although this might be the case, the flat shape was also a most effective antipredator device. Sand dollars are the only group of urchins in these rocks that appear to have been spared the attention of the cassids. The common genus is *Monostychia*. Of the hundreds of specimens that I have examined, only one has a cassid-cut hole in it. The reason for the unpalatability of sand dollars may lie in their internal construction. Within the test are a series of internal supporting struts. Any cassid that managed to bore into the test would immediately be faced with these obstacles to bore through. So the energy invested in trying to bore through to the tasty gonads was probably not recouped even after a good feed.

The never-ending arms race between urchins and cassids has seen the evolutionary histories of the two groups inexorably linked, to the benefit of both groups. Furthermore, it has contributed to the wide diversity of urchin types and has resulted in their evolution into a wide range of marine environments. And this diversity has itself led to the evolution of a wide range of cassids.

Overall trend in sea urchin evolution was to burrow deeper in finer sediments in deeper water. This was probably driven by high levels of predation in shallower water.

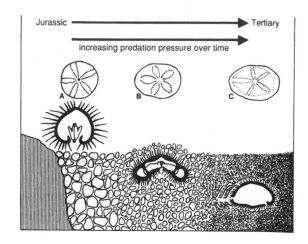

The disappearance of Paleozoic sea urchins, which possessed flexible tests, and their replacement mainly by forms with rigid tests may well also have been partly influenced by predation pressure. Similarly, the flowering since post-Paleozoic times in diversity of urchins, gastropods, and bivalves that burrow within the sediment has been shown to correspond with an increase in diversity of predatory gastropods. Adopting a cryptic habit, by deep burrowing and by habitat displacement into deeper water niches, in a number of urchins may have arisen as a direct response to the adaptive radiation of predatory gastropods. While the high levels of morphological diversity evolved by irregular urchins during the Cenozoic has been attributed to the dominant effect of heterochrony, extrinsic, as well as intrinsic, factors must have played a part. The evidence suggests that predation pressure was the key extrinsic factor.

Predation pressure can work in different ways in different environments and can be very specific in the structures that are targeted, and so which are preferentially selected. A classic example occurs in the three-spined stickleback fish *Gasterosteus*. This fish lives in both marine and freshwater habitats. Usually no more than 75 mm long, the fish bears three dorsal spines, a row of lateral bony plates, and a stout pelvic girdle, all anatomical features that assist in deterring predators. Some species of *Gasterosteus* have invaded freshwater habitats, where distinct morphological differentiation has occurred. One of the most effective defensive structures against vertebrate predators in marine populations is a fully formed pelvic girdle, which bears a prominent spine, combined with a stout posterior process. During the ontogeny of marine forms of *Gasterosteus*, the spine develops in intensity. However, in some freshwater forms paedomorphosis has occurred, resulting in the loss of the defensive spine and reduction of the stout pelvic girdle. Michael Bell of the State University of New York at Stony Brook has interpreted this change as a response to a change in selection pressure.[3] In the sea, the main predators are vertebrates, notably other species of fish. The presence of pelvic spines and processes is a very effective defense against other fishes that try to swallow the sticklebacks. Consequently, there is strong selection pressure for the presence of a robust pelvic girdle.

However, in freshwater habitats predation pressure has shifted away from vertebrates toward invertebrates, mainly insects. Although the pelvic spines and process are effective deterrents against vertebrate predators, in freshwater habitats the selfsame structures become a distinct disadvantage because they are very useful to predatory insects. They grab hold of the spines and use them to grasp the fish tightly. As a result, in freshwater habitats, where predatory fishes are absent but predatory insects are present, selection pressure has favored the evolution of forms of *Gasterosteus* that show a paedomorphic reduction in the degree of pelvic development.

The arms race between the sticklebacks and their predators has a long history. Bell has studied a 110,000-year sequence of Miocene deposits in which fossil specimens of *Gasterosteus doryssus* show paedomorphic pelvic reduction. Through the sequence Bell observed an intraspecific paedomorphocline where, as sticklebacks with a fully formed pelvis declined, intermediate forms with vestiges only appeared, followed by specimens in which the pelvic structure was lost altogether. The absence of other fishes from these horizons where fish with a paedomorphic pelvis occur is used by Bell as an argument for increased selection pressure against those fish with a fully formed pelvis due to high levels of insect predation.

Some animals change the habitats in which they live during their life. They often do this either to avoid predators or to harvest resources from different types of habitats. The time that the shift from one strategy or habitat type to another takes place is often directly related to the animal's size and its rate of growth. During an individual's ontogeny, its food or habitat use may change continuously, or there may be more abrupt, episodic changes (such as the aquatic to terrestrial habitat shift in amphibians). The time at which these ontogenetic habitat shifts occur will be critical to the organisms' life history strategies, especially in terms of exposure to different predators.

The bluegill sunfish (*Lepomis macrochirus*) passes through a number of habitat shifts as it grows up, moving from living among the vegetation in very shallow water at the lake's edge to free-swimming, open-water habitats in lakes, several times during its life history.[4] After they hatch in the near-shore zone, the fish migrate to open water to feed on zooplankton. They then return to the shallow water for some years to feed on invertebrates. When they reach a larger body size they may then once again migrate back to the open water. However, the intensity of predation on the bluegill by another fish, the largemouth bass, affects the body size at which habitat shifts take place. The density of bass in different lakes is positively correlated with the size at which the habitat shifts occur: the higher the density of the predator, the larger the size at which the switch occurs.

Predation risk for bluegills is much higher in open water. On the other hand, high density of vegetation in the near-shore, shallow water zone results in very low levels of predation, the bluegills being able to hide themselves away among the dense aquatic vegetation. Even though the open water is a more hostile habitat, it is energetically mandatory that the bluegills migrate to this more dangerous environment at some stage in their lives so as to consume enough food

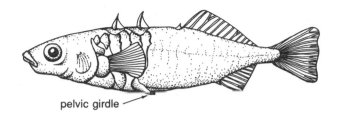

The three-spined stickleback *Gasterosteus* has a prominent pelvic girdle, which helped adapt it for life in open marine conditions.

pelvic girdle

to maintain their larger body size. A strong element of trade-off between the risk of being eaten, against getting enough to eat, comes into play. While the risk of being eaten is reduced as the bluegills grow bigger, there is greater selective advantage for the switch to take place at a larger body size. This means that the timing of habitat switches that are related to body size depends almost entirely upon the presence of the major predator, the largemouth bass. The presence of this fish effectively confines small bluegills to the vegetation-rich shallow water zone. Another effect of higher predator density in open water is to cause the bluegills to spend a longer period in the shallow water zone. Thus, in two lakes where levels of the predator are very different, body sizes at time of habitat shift vary greatly. In one, the bluegills that undergo the habitat shift when about five centimeters long spend only two years there, whereas those in a lake where predator densities are higher do not shift until they have reached a body size a little over eight centimeters long, after having spent about four years there. This itself has a direct impact on growth rates, for as more age classes become restricted to the vegetation in the shallow water, the population density increases, resulting in a lowering of growth rates. This causes the time that it takes for the bluegill to reach an optimum size for a habitat shift to be further extended.

Predation Pressure Affecting Time of Maturation

Whereas Peter Pan seems to have spent much of his extended juvenile life fighting with Hook to stay alive, for many organisms one way to minimize predation pressure and stay alive is to alter the time that it takes to reach maturity. This is because many organisms, like the bluegills, will be subjected to varying levels of predation at different stages in their life histories. So, if the juveniles live in a different habitat from that occupied by the adult, and where predation pressure is much less than is experienced by the adults, selection will tend to favor individuals that spend less time in this more vulnerable adult period and remain longer as juveniles. There are two different ways in which this can be achieved: either by delaying the onset of maturity (hypermorphosis), so extending the juvenile phase at the expense of the adult phase, or by foreshortening the adult phase. The short-lived mayflies have spent so much of their lives as juveniles that there is insufficient time for certain adult morphological features to develop. As a consequence they lack functional mouthparts and have a digestive system that largely fails to develop. In these insects, a relative prolongation of the juvenile phase takes place, at the expense of the adult phase. Time of onset of maturity has been delayed, relative to its ancestor, but the total life span has not changed. Since mayflies evolved during the Devonian Period some 350 million years ago, a dominant trend in mayfly evolution has been for this stretching out of the juvenile period of growth. This has occurred entirely at the expense of the adult phase, which has undergone no corresponding extension. So, like the nonparasitic lampreys (see Chap. 10), the last, adult phase is almost squeezed out of the life cycle altogether. And this has occurred in response to intense predation pressure on the adults.

In many families of mayflies, the adult phase has been ruthlessly decimated to end up being a minute proportion of the entire life span of the animal, often occupying less than two hours out of a total life span of two years. In the case of one species of mayfly, *Dolania americana*, males exist as adults for about one hour, whereas the females attain adulthood for a mere half an hour, out of a life span of fourteen months.[5] This means that the adult phase occupies a mere one ten-thousandth of the entire life span of the organism. If this occurred in humans who reached maturity on their fourteenth birthdays, we would promptly die at noon on the same day!

As juveniles, mayflies are aquatic and predatory. However, when they become adults the tables are turned. During mass matings of mayflies during the dawn hours, following metamorphosis, both the males and females are subjected to extremely high levels of predation by a wide range of animals: birds, bats, insects, and spiders. Predation takes place both in the air and on the water surface where beetles attack the floundering newly emerged adults. Thus, the very intense predation pressure, which is focused only on the adult phase of these mayflies, has led to a strong reduction in the adult phase—like lampreys, by sequential hypermorphosis.

So, body size clearly has a strong effect on the extent to which an organism will suffer predation. Predators operate under what is known as an "optimum foraging strategy."[6] This means that predators will target prey within a certain size range. The outcome of this is that it effectively reduces predation pressure on animals with small and large bodies, most intense predation occurring somewhere in between. In the case of small-bodied prey, the energy expended by predators when they attack small prey does not produce a sufficiently high return on the energy investment; if prey body size is too large, the predator has to expend too much energy for an adequate return on its investment. There are also the basic physical problems of trying to tackle large prey. When predation pressure is being exerted, selection will favor forms that reproduce earlier and at a small size or later at a much larger size. There are also advantages in rapidly attaining a large body size by an acceleration in growth rate, both for the predator and for the prey. It takes the prey out of the range of the optimum prey size quicker, while it allows the predators to get into their predatory range earlier.

For instance, in the little bivalve *Nucula turgida*, which lives in shallow areas of Dublin Bay in Ireland, only shells of a certain size are attacked. While this bivalve only rarely grows any longer than ten millimeters, boring gastropods that prey upon them rarely take the smaller-sized individuals.[7] Because the bivalve reproduces most at shell lengths between five and seven millimeters, when at three to five years of age, predation on larger shells reduces competition among the prey, favoring the smaller forms that have greatest reproductive potential. Selection pressure has therefore favored forms that can reproduce relatively early, the upshot being the evolution of a species with a relatively small shell.

So, in cases from sea urchins to fish to bivalves, we can see that varying levels of predation can have a strong influence on natural selection. For whatever survives is being preferentially selected, because, perhaps, of some particular attribute that it possesses, be it anatomical, ecological, or behavioral. High levels of predation can actually have a direct effect on causing populations of organisms to change their growth rates or maturation times and so indirectly influence the size and shape of the adult organism.

This has been shown experimentally in the water flea *Daphnia magna*, collected from seven sites in eastern England. In these experiments, the effect of a predator was imitated by culling. After having divided the entire size range into 11 classes, the lower size classes, 0 to 6, were repeatedly removed from one group and the upper size range, 7 to 11, from another. While culling the lower size range revealed no significant change over the time that the experiment was being carried out, removal of the large sizes resulted in a steady decline in numbers in this category being produced by the organisms in subsequent generations. Somehow the extent of ontogenetic development was essentially being influenced in descendant populations by what was happening to their ancestors. They were effectively bequeathing a greater chance of survival to their as-yet-unborn descendants.

When the culling period had come to an end, the structures of the populations had significantly changed, because both growth rates and the timing of onset of reproduction had changed. After large sizes were excised from one population over a long period, the clones (for reproduction in these water fleas was by parthenogenesis) grew more slowly. Moreover, they reproduced and died at a smaller size than clones from other treatments. They also reproduced at an earlier age. So, following culling of large sizes, the clones that were left grew more slowly through their early growth stages, whereas in treatments where smaller animals had been preferentially culled, the clones that remained grew rapidly through this early growth phase, about twice as fast as the rate for those of the large culled treatment.

What all this artificial predation did was to result in the selection of clones that dwelt in the more vulnerable size ranges for as short a time as possible—they shot through as quickly as they could. Where culling targeted the small size classes, selected clones grew through this phase more quickly; where large sizes had been culled, clones took much longer to reach this more vulnerable stage—they were in no rush to get into what was to be a more vulnerable stage in their life history. The conclusion reached is that size-specific culling seems to result in genetic differentiation in rates of growth and maturation times. Escaping from the more vulnerable periods of predation when the fleas are most likely to survive is achieved by heterochrony.[8]

Perhaps the most stunning example of predators directly influencing the developmental history of their prey is shown by a little freshwater snail, *Physella virgata*, which lives in spring-fed streams in Oklahoma. For here it is not just natural selection favoring forms with some particular propitious growth rate or maturation time, but a direct biochemical assault on the prey that induces changes in the snail's hormonal system, and thus its life history. Experiments carried out by Todd Crowl and Alan Covich at the University of Oklahoma have revealed that the size to which the shells grow (which is determined by the time that it takes them to reach maturity) and their length of life can be directly influenced by the presence or absence of predatory crayfish. The snail is characterized by its short life cycle, young age at maturity, and short reproductive period. Those populations that live free from the activity of the predator show rapid growth until the shell is about 4 mm long, at which time reproduction begins and the rate of growth declines markedly. These snails live for three to five months. In their natural environment the snails are preyed upon by a number of different types of animals: fish, insects, birds, and crayfish. If crayfish are living in the streams occupied by the snails, then the snails delay their onset of maturity, and so grow to a larger size (up to 10 mm long). As a result, individuals live longer (eleven to fourteen months).

To understand why this should be so, Crowl and Covich experimented by taking water from a stream in which only snails lived, and water from another stream in which both snails and crayfish occurred together. They then raised two separate populations of snails in these different waters and found that even though the crayfish weren't lurking in the water, those snails living in the water from streams in which crayfish occurred lived longer and grew larger than those snails raised in the crayfish-free water. To Crowl and Covich this was firm evidence that a chemical cue had been introduced into the water by the crayfish. This was detected by the snail, and had the result of delaying the onset of sexual maturity. The effect was that these snails stayed in the juvenile high growth rate period for longer and attained a larger size before they reproduced. The crayfish are very finicky feeders. They will only go for small snails. Anything bigger than 10 mm long they won't touch. It is therefore to the snails' advantage to grow to a larger size as rapidly as possible. This switch in the time that the snail achieves maturity caused by the chemical cue introduced by the crayfish must act directly on the snails' hormonal system.[9]

Predation Driving Macro-evolution

Sitting back and thinking about the major steps in evolution that have taken place over the last half a billion years, the nagging questions that usually remain unanswered are, What caused these major advances to occur when they did? Why did evolution proceed in the direction that it did? Exactly what was the underlying impetus driving evolution in any one particular direction? The classical Darwinian struggle for survival, and "may the best individual win," is not only a description of competition between individuals of a species, of one bag of genes trying to outdo another, but also of the classic "survival of the fittest." Surviving long enough to reproduce is a mighty battle for many individuals of most species. Just think of the hundreds of eggs laid by a single turtle, and the tiny number of offspring that survive long enough to reproduce. The same goes for so many invertebrates. Of the baby spiders that I talked about in Chapter 5, only a minute proportion will have survived. The two factors that conspire to reduce the population drastically are either lack of food or ending up as someone else's food. For a potential prey species (and this makes up most species), the aim of the game of life is to outwit, outfox, or outrun the predator. The longer you can avoid becoming something else's meal, the greater the chance of passing on your genes to another generation.

It has been argued that the Cambrian Explosion, that stunning episode in Earth's history (a period perhaps as short as ten million years) when all the basic body plans of the major phyla that we see today sprang forth (along with a few that have fallen by the evolutionary wayside), was driven by predation pressure. Just consider what actually happened at that time. One of the most significant events of this evolutionary explosion is that almost geologically simultaneously a wide range of organisms acquired the ability to secrete a hard outer covering—a shell, which served as armor. Even in some of the oldest of these deposits, like the amazing Chenjiang fossil site near Kunming in Yunnan in southern China, huge predators like *Anomalocaris* were present. Up to a meter in length and armed with a vicious set of "teeth" set in a circular jaw that resembled a pineapple ring but worked like a dagger-lined vise, they were the kings of the seas. Trilobites with obvious bite marks provide direct evidence that even with an outer shell, predation was an everyday occurrence. It has been argued that the shape of some bite marks fit perfectly the "mouth" of *Anomalocaris*. We see during early Cambrian times a great proliferation in types of trilobites. There is appreciable variation in sizes (from a few millimeters to about half a meter long) and in numbers of thoracic segments (from two to more than fifty). The extent of regulation of genes that control development was probably relatively poorly constrained. The result was an evolutionary explosion as a myriad of forms evolved, providing a potpourri of items for natural selection to work on.

Quite a few examples of progenesis have been documented in Cambrian trilobites.[10] The little trilobites I found in my quarry in northwest Scotland, which look for all the world like overgrown larvae, but for all that are still tiny adult trilobites, may not have evolved because of their particular morphological attributes. It could be argued that because precocious maturation led to a free-swimming life, it may have been advantageous in an environment where many new predators were coming on the scene to mature earlier. Life in the fast lane occurred not by choice but by the necessity to survive to adulthood and reproduce. The startling "windows" afforded by deposits like those at Chenjiang and the Burgess Shale in British Columbia, through which we can glimpse a time long past, more than half a billion years ago, reveal a veritable menagerie of arthropods. And some of these are likely to have been predators.

Many of the earliest known animals that crawled along the Cambrian sea floor were heavily armored. And many of these tended to be quite small. The enigmatic little tommotiids, looking, as Simon Conway Morris of the University of Cambridge puts it, "broadly similar to an armoured slug"; the equally enigmatic wiwaxiids, either mollusks or polychaete worms, depending on your point of view, armored with a covering of what looks like rows of razor wire interspersed with steak knives; and the wormlike halkieriids, with their covering of chain mail and broad shields at head and tail, all point to an evolutionary imperative to protect themselves from other, bigger, and not so friendly contemporaries. As Conway Morris points out, the Middle Cambrian Burgess Shale contains a surprising abundance of predators and scavengers. As he argues, the form of the mouth parts and the nature of the gut contents are compelling evidence for the existence of predators at this time. Spinose appendages in arthropods point to this, as do the crunched-up brachiopods and the conical-shelled hyoliths in the gut contents of the large arthropod *Sidneyia*. Conway Morris points out as well that the priapulid worm *Ottoia* also ingested whole hyoliths and, furthermore, appears to have been cannibalistic.[11]

Another evolutionary leap of some magnitude was the colonization of land by animals, first by arthropods, then by tetrapods, during the Early Paleozoic. Indirect evidence from trackways points to the first tentative steps occurring at least 430 million years ago. As I outlined in Chapter 8, many of these tracks were probably made by large, predatory arthropods, such as the giant scorpionlike eurypterids. Of the actual body fossils, the earliest occur in 415-million-year-old siliceous rocks in Shropshire in England. Here occur the beautifully preserved remains of centipedes, spiderlike trigonotarbids, and mites. Other slightly younger sites in New York State show a similar picture of predatory and scavenging arthropods eking out an existence on an inhospitable land. Plant cover would have been minimal, so there would have been a lack of well-developed soils, hence wind erosion; and sandstorms were probably a common occurrence. If the earliest land arthropods that existed in rivers and pools were capable of an amphibious existence, then the longer they could persist on land out of pools of water in which many predators lived, the better their chance of survival. As I have shown, the earliest insects were likely to have been very small. Maybe their evolution had little to do with the economy of generating just three pairs of legs. Perhaps it was the small body size that allowed them either to hide more effectively from predators or just to be out of the range of the feeding strategies of many of the larger arthropods.

So often interpretations of the evolution of organisms into new ecological niches suffer from what I like to call the "Star Trek Syndrome": as though the species set out "to boldly evolve where no species has evolved before." In other words, if there is a vacant ecological

Evolution in
the Early
Cambrian of
protective
armor: *Hal-
kieriia (right)*;
Wiwaxia (left)

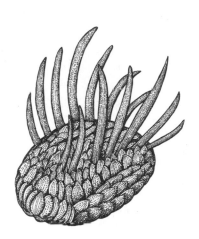

niche, it must be filled. I think it is much more likely that rather than just a passive drift into major new niches, it occurred by a driving force, in the form of predation pressure. In the case of the early arthropods and the first tetrapods, which followed soon after, it was one thing for there to be a vacant niche in which to evolve; it was another to evolve the necessary morphological and physiological adaptations to cope with the effects of gravity, low oxygen levels, large diurnal temperature ranges, and the ever-present danger of desiccation. Even if the requisite morphological and physiological problems had been overcome, and the environment was suitable, there needed to be an external selection pressure to force animals into this new, and very different, niche.

The first appearance of tetrapods on land, nearly 400 million years ago, may likewise have been predator-driven. As well as a proliferation in the aquatic environment of arthropod predators, such as eurypterids, and large molluskan predators, like cephalopods, there was a great increase in the diversity of fishes around this time, providing yet another source of predation pressure. Any bizarre variety of fish capable of existing out of water, even if only for a short time, is likely to have had a better chance of winning the lottery of life.

In earlier chapters I outlined the important part played by heterochrony in the evolution of flight in insects, reptiles, birds, and mammals. Though we can never be sure, it is not unreasonable to suggest that the most likely reason for these animals to have evolved into this major new, aerial niche was the impact of predation. If you were a small theropod dinosaur, it would be useful to have some sort of means of avoiding being eaten by a larger theropod. What better way than to be able to escape into the air and look down and blow a raspberry at the hulking brute whose snapping jaws were catching nothing but a mouthful of air? For groups like the ratites, the subsequent loss of the ability to fly in these giant flightless birds was more than compensated for by the large body and powerful legs that would deter, for the most part, the largest of predators. But in New Zealand and Madagascar, it was the brain of the predator, not its brawn, that was the flightless birds' nemesis, because humans hunted them to extinction.

The impetus for the evolution of groups like burrowing irregular sea urchins may have been the evolution of a wide range of predatory gastropods during the Mesozoic. The evolution into this cryptic niche was permitted by the heterochronic changes in growth patterns in Jurassic urchins that saw a freeing up of the architectural constraints that had hitherto restricted the range of morphologies that could evolve. As a group, urchins display many of the different types of antipredation strategies that organisms have evolved: habitat selection strategy, such as evolving into areas like deep water, where there are fewer predators; fortification strategy, wherein improved defensive structures, like spines, evolve; crypsis strategy, in which the organism hides from potential predators, either by burrowing or by developing effective camouflage; and chemical-warfare strategy, evolved by some urchins that emit a powerful chemical that deters potential predators.

The ability of organisms to evolve morphological traits by heterochrony allows the occupation of new niches or new strategies to minimize predation pressure. If morphological traits that permit the occupation of new niches do not evolve, or if suitable niches are not present, then extinction ensues. So long as the prey species keeps one step (or one niche) ahead of the predator, then it can survive. Once this advantage is lost, the grim reaper wields his glistening scythe and consigns the species to eternal extinction.

12

The Baby-faced Super Ape

"A foetal ape that's had time to grow up," Dr. Obispo managed
at last to say. "It's *too* good!" Laughter overtook him again.
"Just look at his face!" he gasped, and pointed through the bars.
Above the matted hair that concealed the jaws and cheeks,
blue eyes stared out of cavernous sockets. There were no
eyebrows; but under the dirty, wrinkled skin of the forehead,
a great ridge of bone projected like a shelf. . . .

   Dr. Obispo went on talking. Slowing up of development
rates . . . one of the mechanisms of evolution . . . the older an
anthropoid, the stupider . . . senility and sterol poisoning . . .
the intestinal flora of the carp . . . the Fifth Earl had anticipated
his own discovery . . . no sterol poisoning, no senility . . . no
death, perhaps, except through an accident . . . but meanwhile
the foetal anthropoid was able to come to maturity. . . . It was
the finest joke he had ever known.

   Without moving from where he was sitting, the Fifth Earl
urinated on the floor.

Aldous Huxley, *After Many a Summer Dies the Swan*

WITHIN ALDOUS HUXLEY'S DESCRIPTION OF THE FINAL FATE OF THE FIC-
titious Fifth Earl of Gonister lie what have been articulated at various
times, and to varying degrees, as the two fundamental components
that have driven human evolution. One is the "slowing up of developmental rates," in other words, neoteny resulting in paedomorphosis.
If we are to accept this interpretation, then it means that we are little

more than apes that have failed to develop. The other is "the older an anthropoid, the stupider," the acceptance of the fact that compared with every other ape species still in existence, and as far as we know every ape species that has ever existed, we live much longer. Huxley was writing at a time when paedomorphosis seemed to explain everything. So if humans had attained their paedomorphic state by living longer and slowing their developmental rate, then the logical conclusion would be that if our ontogeny was stretched out even further, we would become even more juvenile and regress to become little more than overgrown fetuses. But has human evolution really been that simple? And are we really that stupid? Like the rest of the animal and plant kingdoms, in our evolutionary journey, we too, as a species, have been molded by variations in the rates and timing of development of our ancestors. That much is clear. But has the driving force, as many have argued, been paedomorphosis, producing younger and younger features? Or have we developed in the opposite direction and become peramorphic, delaying our maturity and developing "beyond" our ancestors? Then again, could we be, like so many other species, a unique mosaic of characters, some speeded up, others slowed down, producing one of the more potent cocktails that three and a half billion years of evolution has managed to concoct?

So, Where Do We Fit into All This?

Considering the relatively peaceful acquiescence among biologists and paleontologists during the last two decades that there are, in most cases, equal opportunities for both paedomorphosis and peramorphosis, the somewhat heated discussions that have taken place concerning where *Homo sapiens* fits into all of this are, in some respects, surprising. I suppose that, as evolutionary biologists, we are happy to sit back and discourse about whether it was this mechanism or that, which crafted the particular species we are studying. But when it comes to the species closest to our hearts, we have been much more circumspect. I know that by giving over a whole chapter in this book to looking at the role of heterochrony in one species I could be accused of being a trifle biased. But if, as I have argued, changes in development have played such a crucial role in evolution, then there is no reason why they should not also have contributed to our own evolution. Indeed, I would argue that to understand fully the evolutionary history of our own species, we must take into account how the developmental programs of hominids have changed over the last four million years or so. For in many ways the influence of heterochrony in fashioning our species encapsulates so many of the factors that I have talked about in earlier chapters that it provides one of the better examples of how important changes in the developmental program have been to our evolution.

Despite the neglect that heterochrony has suffered in studies of evolution, there has, since early last century, been some interest in its role in human evolution. Views have been polarized, however, some arguing from a paedomorphic perspective, others from a peramorphic one. To appreciate current perspectives, it is useful to review how the changing attitudes toward heterochrony in evolution over nearly two hundred years have colored our views of how human anatomy has evolved. With such changeable fashions, the "emperor's new clothes" factor comes into play. Those on either side can see some morphological or developmental characteristics that support their argument, while opposing ones seemingly just don't exist; what is plain as the nose on your face to one person will be invisible to another. We see what we want to see.

Are We Really Just a Babe among the Apes?

Even though "recapitulation" held such a stranglehold last century, a number of interesting observations were made noting the surprising similarity between some adult human anatomical features and those of juvenile apes. The most striking of these is the strong similarity between the appearance of the head of a juvenile ape and that of an adult human. As Stephen Gould relates in *Ontogeny and Phylogeny*, as early as the 1830s the French transcendental morphologist Etienne Geoffroy Saint-Hilaire was struck by the resemblance between juvenile orangutans and humans. He observed that "In the head of the young orang we find the childlike and gracious features of man," and contrasted this to the head of the adult orang, noting that "if we consider the skull of the adult, we find truly frightening features of a revolting bestiality."[1]

Geoffroy Saint-Hilaire and his transcendentalist followers believed in the concept of a "chain of being," in which all animals essentially follow a single, structural plan. This was firmly encased in recapitulation. So, to Geoffroy Saint-Hilaire there can have been no other explanation than that the orang was simply an anomaly. Indeed, as Gould has argued, this same attitude persisted throughout the nineteenth century whenever any other similar cases of apparent similarity between adults and the juveniles of other species lower down the chain were observed. After all, was not *Homo sapiens* the pinnacle of creation? How could we be *less* developed than an ape? The very idea was, to any right-minded person in the nineteenth century, preposterous. As our old friend Edward Drinkwater Cope saw it, along with so many of his contemporaries, "man [stood] at the summit of the Vertebrata." However, Cope had to admit that in certain "extremities and dentition," humans seemed to resemble "the immature stages of those mammals which have undergone special modifications of limbs and extremities." In a similar fashion Cope accepted that the shape of the head (in particular the protruberant forehead, vertical face, and reduced jaws) must have been products of "retardation." Yet, concerning the "important" features (the nervous and circulatory systems and much of the reproductive system) Cope argued that these were a product of "acceleration," and so could slot nicely into the recapitulation model.[2]

However, once the biogenetic law nosedived into disrepute, to be replaced by paedomorphosis as the all-important driving force in evolution, there was no holding back those who argued forceably for a paedomorphic origin for humans. Principal protagonist among these was Louis Bolk, professor of human anatomy at Amsterdam, whose concept of "fetalization" has been championed in more recent times by Gould. Bolk considered that a number of anatomical features found in adult humans were akin to the features present only in juvenile apes. Indeed, Bolk believed that what distinguishes us from all other primates were these "fetal conditions that have become permanent." Among these were

1. our flat face;
2. our generally hairless state, compared with other primates;
3. loss of pigmentation in the skin, eyes, and hair;
4. the form of the external ear;
5. the central position of the foramen magnum in our skull;

6. the relatively high brain weight;
7. the form of the hand and foot;
8. the structure of the pelvis.[3]

More recently, followers of the paedomorphic school of human evolution, in particular Ashley Montagu in the 1980s, included yet more characteristics of adult humans that they perceived to be paedomorphic. These include

1. absence of brow ridges;
2. thin skull bones;
3. small teeth;
4. teeth that erupt relatively late;
5. a prolonged period of infantile dependency;
6. a prolonged period of growth;
7. a long life span;
8. a large body size.[4]

These last four features have, I believe, been particularly greatly misrepresented, misinterpreted, and misunderstood. They are in fact anything but paedomorphic. Yet, within them may lie the key to unraveling the fundamental relationship between our ontogenetic development and our evolutionary history, for the two are inexorably linked.

Bolk argued that "fetal" growth rates arose from "delayed development," producing "retardation": around this premise was based his entire thesis. He viewed humans as growing very slowly, compared with other primates, taking much longer to attain their final form. If this were so, all these features that are purported to be paedomorphic would be the product of neoteny. Bolk's view of life's course as progressing at a very slow tempo can easily be tested by comparing us with nonhuman primates. Brian Shea of Northwestern University is one who has argued strongly against paedomorphosis as being the main driving force in human evolution. He has pointed out, as have others, like Kathleen Gibson of the University of Texas at Houston, that just a simple comparison with the growth rates of the chimpanzee (which, after all, is genetically almost 99 percent the same as us) shows there to be no difference at all.[5] We do *not* grow any slower. The only thing that is "slowed down," if it can be expressed like that, is the transition from one anatomical stage to another as ontogeny progresses, including aspects such as the time that we become sexually mature and the accompanying reduction, then cessation, of growth.

Much of the confusion that has arisen concerning the role of heterochrony in human evolution has done so for more than one reason. Not only do we desire to distance ourselves as much as possible from any possible connotations of recapitulation (and the resultant assumption that we must therefore be the most complex and therefore the best), but I believe there has been a basic mistake in equating delays in transition from one growth phase to another with reduction in growth rate. They are nothing of the sort. Had humans been the product of reduced, neotenic growth we would be vastly different beasts, small of stature, small of limb, and, significantly, small of brain. Delays in transition from one growth phase to the next and neoteny are empirically different processes, yielding fundamentally different results. In the case of humans the product of our pattern of development, which is characterized by long, drawn-out growth phases, is overwhelmingly not one of paedomorphosis but one domi-

Differences in proportionate growth in humans and chimpanzees. Reduced to the same sitting height.

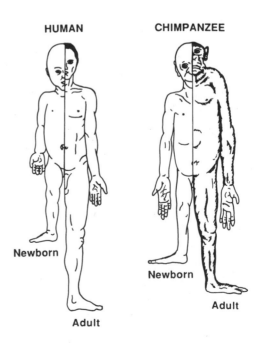

HUMAN

CHIMPANZEE

Newborn

Adult

Newborn

Adult

nated by peramorphosis. In many important ways we have developed "beyond" our ancestors and all other primates. However, fear not. There is no need to raise the specter of recapitulation, for, as I will show, the peramorphic nature of our species is the product of much more than simple terminal addition to our ontogenetic development.

In his landmark book *Ontogeny and Phylogeny*, Gould argued strongly that humans are paedomorphic: "human beings are 'essentially' neotenous, not because I can enumerate a list of important paedomorphic features, but because a *general temporal retardation of development has clearly characterized human evolution*" (Gould's italics). In this view, therefore, even the delay in onset of changes from one state to another (such as the time of eruption of teeth), and in onset of maturity itself, are regarded as retardation and therefore, *ipso facto*, paedomorphosis. Yet the modern synthesis of heterochrony, which applies to the rest of the animal and plant kingdoms, can accommodate the changes quite easily in other distinct mechanisms, like hypermorphosis, or postdisplacement, the former producing extended, peramorphic, not paedomorphic, evolution.

So, what about our "baby faces" and apparent juvenile ape appearance? A much reproduced figure is one published by Gould that shows a juvenile chimp alongside an adult. While I would agree that most of my own acquaintances bear slightly more of a resemblance to the juvenile in overall head shape than they do to the adult, does this single factor have the right to color our entire view of human evolution, as it seems to have done? Does it really mean that we are apes that have failed to grow up? Kathleen Gibson, who has worked extensively on chimpanzee development, pointed out to me that the upright, almost noble profile of the juvenile chimp depicted in this figure is a most misleading and totally unnatural pose. The adult, on the other hand, is shown in typical, hunched, apelike pose, accentuating the apparent differences between the juvenile and adult. As I have argued in earlier chapters, few, if any species are entirely paedomorphic or entirely peramorphic—they are a cocktail of features, some speeded up, some slowed down. And so it is with humans.

What needs resolving is the relative importance of the two. When the shape of our head is compared with that of our ancestral hominids, it can be seen that there has been a change in relative proportions of the skull. In very general terms, there has been a relative reduction in the size of the jaw, when compared with the skull. The skull itself has undergone more actual growth in *Homo sapiens*, compared with any other primate, whereas the extent of jaw development has not kept pace with brain expansion. As I shall elaborate below, the increase in brain size, and thus skull size, is a logical outcome of our extended period of growth—what is essentially our hypermorphic development—compared with other primates. The jaw, however, has dissociated its growth and experienced very little relative increase during ontogeny, compared with other primates. Thus, we share with ostriches and the tyrannosaurids a case of developmental trade-off. The brain size has increased by leaps and bounds. It has been at the expense of the jaw, and, as I shall argue below, of other organs. The result of this is the overall similarity between a juvenile chimp head and an adult human's; but it arises from a complex mosaic of different heterochronic strategies operating on different parts of the head. It is not simply a case of neotenic reduction in the "amount" of growth.

It is curious that those who still argue for a paedomorphic origin for humans view the prolongation of early juvenile growth rates as producing paedomorphosis, because the juvenile growth rates are continued for a longer time to a later stage. The outcome of this may still be a larger, more complex structure, which in this later stage of development bears little resemblance to the same feature in the ancestral juvenile; but by an odd twist of logic, this is still perceived as being paedomorphosis. So, attributes like a delay in the closure of the sutures of the skull (these allow the skull to attain a much larger size) are described as "neoteny"! Yet such delays in transition from one stage to another (open to closed in this case) are not "rate" features at all. They are the much more significant, but largely ignored, aspect of human evolution—sequential hypermorphosis—the progressive delay in transition from one growth phase to the next, producing a stretching out of the whole period of development. This is not simply a case of semantics, for sequential hypermorphosis produces fundamentally different outcomes, resulting in a range of important anatomical, physiological, and behavioral characteristics that are in no way juvenile ape characters. On the contrary, they have resulted in our developing well beyond any ape that lived in the past or that is living today.

Let us look at another so-called paedomorphic characteristic on

Bolk's list—the form of the human foot. Bolk would argue for this being a typical paedomorphic trait. After all, the foot of an embryonic human, when compared with that of an embryonic ape, is very similar. As each species grows and develops (the human for much longer), the toes of the ape become relatively longer than do those in the human (with the notable exception of the big toe, which grows at a faster rate). Shorter toes must therefore mean a paedomorphic feature, and therefore human feet are quite clearly paedomorphic—Q.E.D. However, the growth of any anatomical features cannot be viewed in isolation from neighboring elements. The foot, after all, consists of other bones that make up the foot proper, including the heel. These grow at much a greater relative rate during human development than they do in nonhuman primates. This produces a relatively larger foot surface, capable of supporting our greater body weight. Once again there is developmental trade-off. While longer toes will be more useful when climbing, the relatively larger, peramorphic foot is better for walking and supporting the body weight of a primate that can walk upright on its two feet.

Another supposed paedomorphic feature is what is termed the "basicranial flexion." This is the angle that the head makes with the body. Work on embryonic and young skulls of apes shows that there is a distinct difference in the state in humans, despite the superficial resemblance between adult human basicranial flexion and the situation in nonhuman primates. This arises, as Brian Shea suggests, from humans following a quite different developmental pathway from other primates. This convergence is a result of a functional need to increase the supralaryngeal space arising from the pronounced peramorphic postnatal growth of the tongue and larynx. These undergo much more growth during ontogeny in humans than in other primates, permitting novel modification of sound, and hence speech and language.

Similarly, our apparent juvenile primate face is *not* due to simple retention of juvenile growth patterns. Studies of the evolution of the human face have shown that when facial and mandibular growth of modern humans, chimpanzees, early hominids (in particular *Australopithecus africanus*), and other australopithecines are compared, modern human patterns of facial growth are a consequence of unique growth patterns that involve, in part, some resorption of bone over

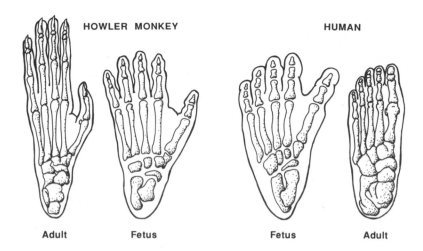

HOWLER MONKEY   HUMAN

Adult   Fetus   Fetus   Adult

Relative growth of the bones of the foot in howler monkeys and humans. While in humans the toes may be considered to be relatively paedomorphic, the bones of the foot undergo greater growth and are peramorphic.

much of the midface and bear little resemblance to growth patterns of other primates.[6]

Lastly, the human brain. It has been asserted that the relative size of the human brain is similar to that of our ancestors' juveniles. However, this is very misleading, for when the development of the brain is viewed in isolation, it can be seen to grow for a longer period than in other primates (hypermorphosis); it attains a larger overall size; and is more complex, in terms of neural proliferation and the development of specialized regions like the cortical areas involved in speech production, information processing, and mental constructional skills. The paedomorphic nature of our head, and of our brain in particular, is a myth. We have the greatest neural information processing capacity of any primate. Certainly, while overall our bodies display a few paedomorphic features—for example, we produce less body hair than other primates (and presumably our ancestors), and we have smaller toes and jaws—these are really not the stuff that great evolutionary breakthroughs are made of. Selection is unlikely to have focused strongly on a naked, short-toed, small-jawed primate. No, the three fundamental aspects of our anatomy that have contributed most to our evolutionary "success" are the fact that we can walk upright, on long, powerful legs, supported by large foot bones that develop more in ontogeny than other primates; the possession of large brains, which allow us to communicate in a complex manner; and, lastly, the fact that we are blessed with an extended juvenile period, during which we are subject to a prolonged period of learning and are able to absorb large quantities of information. Not one of these characteristics is the product of paedomorphosis. On the contrary, they are generated by peramorphic processes that have stretched out our growth phases, more so than in any other primate, and have produced—the super ape.

The Rise of the Super Ape

Our very early ancestors, australopithecine primates who lived up to 4.5 million years ago, were much smaller than we are. The earliest hominid, *Ardipithecus ramidus*, found in 4.3- to 4.5-million-year-old rocks in Ethiopia, is known from just seventeen bits and pieces, mainly cranial material, especially teeth. We know that it had a more primitive tooth morphology than later members of *Australopithecus*. In many ways, including size, it would have resembled a chimpanzee, but its much smaller canine tooth and form of the base of the cranium indicate that it was a hominid. Fragments of arm bones reveal that its body size was as large as some of the smaller, later australopithecines, weighing in at, perhaps, about 30 kg.[7]

Australopithecus afarensis, which lived between 3.4 and 3.8 million years ago, was a little larger, weighing between 30 and 45 kg. From their anatomy and the footprints that a group of three left as they walked across a freshly fallen bed of ash at Laetoli in Tanzania, we know that they didn't slouch like their tree-dwelling ape ancestors— they walked upright, at least for some of the time. Their brain was much smaller than ours, with a volume of about 400 cubic centimeters (ours is nearly 1,400).[8] But just because these early primates were smaller than we are, does it mean that they matured at an earlier age, or were their growth rates less than ours? In addition to fossils of adult australopithecines, fossils of juveniles have also been found. Of critical importance to our understanding of the evolutionary mechanisms within primates are their teeth. Fortunately teeth are very resilient and in vertebrates in general are often the only parts of the body that survive. One of the more significant aspects in terms of their use in deducing patterns and processes of hominid evolution is the relative times of their eruption. From such information we can glean direct evidence of the relative lengths of different growth phases. Like the use of annual growth rings to date trees, teeth can similarly be aged because they lay down daily markings in the enamel. Circaseptan (7–9 day) cycles can also be calculated. From this the age of the individual can be calculated, and thus times of eruption deduced.

As an aid to studying human evolution, teeth are second to none, for they provide distinct markers of life history and allow different developmental stages to be categorized. Using the age at which teeth erupt, hominid (and, indeed, all mammalian) postnatal growth can be classified into distinct phases: infantile, juvenile, and adult. These are based on periods before, during, and after eruption of permanent teeth, respectively. As Holly Smith of the University of Michigan at Ann Arbor has stressed, timing of tooth eruption is critical for animals in terms of their food-processing abilities, and thus growth of the body as a whole. It also corresponds with other life history factors, such as gestation period, timing of onset of sexual maturity, and length of life. Thus, the later the eruption of permanent teeth, the longer the gestation period and the later the weaning, the longer the infantile and juvenile dependency, the later the onset of sexual maturity, the larger the brain size and body size, and the longer the life span. Clearly, because of this, any analysis of the age of tooth eruption in fossil hominids will be very instructive in providing information on all of these life history parameters. Dental development, unlike some life history traits, is relatively insensitive to fluctuating environmental factors. This means that it can act as a good proxy for assessing the timing of sexual maturity and thus enable us to establish heterochronic changes between hominid species.[9]

Even if the actual age of an individual is not known, estimating the time of eruption of the first molar tooth can be reasonably accurately predicted if the brain weight is known because there is a very high correlation between the two. This applies across all primates. So, as Holly Smith points out, the smallest primate, *Cheirogaleus medius*, with a brain that weighs only 180 grams, has the earliest emergence of the first molar of all primates. From this it follows that it has the earliest onset of maturity, smallest body size, and shortest life span. At the other end of the scale comes *Homo sapiens*, which has the largest brain of any primate, correlating with the latest onset of maturity (and therefore the latest eruption of the first molar), largest body size, and longest life span. Using these correlations, estimates can be made of the life history patterns of extinct hominid species, allowing heterochronic patterns to be interpreted.

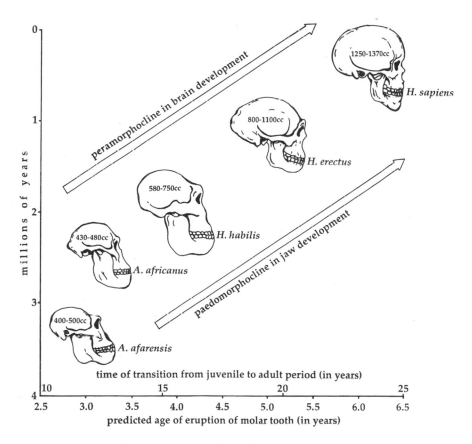

Generalized view of changes in cranial morphology in hominid evolution. Only some species have been selected. As the time of transition from infantile to juvenile growth phases (as measured by time of first molar eruption) and from juvenile to adult periods is delayed, the brain size increases nearly threefold, as does body size. Straightening of the face may be a consequence of the great increase in brain size "pushing" the whole skull forward, dragging the face with it.

From such studies it has been deduced that in early australopithe-cines the first molars erupted when they were between three and three and a half years old, while these early hominids lived until they were thirty-five to forty years old. This is comparable to a chimpanzee. In the later hominids, *Homo habilis* (which appeared about 2 million years ago) and early forms of *Homo erectus* living about 1 million years ago, the first molar erupted when they were between four and four and a half years old. Under favorable conditions these species could have lived until they were about fifty years old. In more recent *Homo erectus*, which lived a few hundred thousand years ago, the first molars would have erupted at about five and a half years of age, while in "modern" humans, who may have appeared as recently as about 100,000 years ago, the first molars erupt after six years, and we have a life span in excess of seventy years. These estimates are made on the basis of increasing brain size found in these hominids, from about 400–500 cc in *Australopithecus afarensis*, increasing to about 430–480 cc in the later *Australopithecus africanus*. The jump to the early species of *Homo* was quite substantial and saw an appreciable in-crease in overall body size around 2 million years ago, when an evo-lutionary burst saw *H. ergaster*, *H. rudolfensis*, and *H. habilis* evolve and *Homo erectus* begin its great march out of Africa. Brain capacity jumped to 580–750 cc in *H. habilis*, increasing to 900–1,100 cc in later *Homo erectus*, finally reaching up to 1,370 cc in modern (that is from about 30,000 years ago to the present day) *Homo sapiens*.[10]

Given this, we can try to estimate life history parameters for the so-called neanderthals, considered by some to be a subspecies of *Homo sapiens*, but by others to be a full-blown species, *Homo neander-thalensis*. Living between about 250,000 and 35,000 years ago, they possessed a brain volume larger than ours, reaching as much as 1,750 cc. Interestingly, the earliest of the "modern" humans in Eu-rope also had a larger brain and larger body. Why modern *Homo sapi-ens* survived while neanderthals didn't, then, becomes quite intrigu-ing. Certainly, in terms of these overall trends in the evolution of life histories, one would have laid bets on *neanderthalensis* winning the evolutionary race, rather than *sapiens*. If hominid evolution can be characterized by this stretching out of development, then compared with us, neanderthals would be expected to have had a larger body size; a longer gestation period; a later transition from the infantile to juvenile phases; a later time of onset of maturation and as a consequence a longer period of parental care of the offspring.

One of the more vexing questions in anthropology is why neanderthals died out. Although they are known to have coexisted in time with modern *Homo sapiens*, there is no direct evidence that the two types lived in the same region and came into contact with each other. As Nancy Minugh-Purvis of the Medical College of Pennsylvania has pointed out to me, while there is evidence in Europe of an intermixing of Châtelperronian and Aurignacian tool industries at some sites (the former probably attributable to neanderthals), it is not clear who made the earliest Aurignacian tools. Moreover, with very low population densities that probably existed at that time, it is not unreasonable to question whether such populations actually coexisted or only occasionally came into contact. So, because life history traits have been so important in human evolution, interpreting them in neanderthals may provide clues to their demise.

The larger body that neanderthals possessed (and, indeed, also the earliest, larger modern *Homo sapiens*) would have required more food to fuel it. Indeed, a large cortex and more sophisticated cognitive development require high levels of energy input.[11] During embryonic development in humans, brain tissue (especially cortical tissue) uses the most metabolic energy. Thus, the evolution of large brains occurs only when the benefits outweigh the costs associated with high-input energy. On top of this, the later onset of sexual maturity, and attendant long childhood, would have necessitated longer parental care. As I discussed in Chapter 11, many organisms live life in the fast lane and try to scoot through their juvenile phase as quickly as possible so as to obviate the vagaries of predation pressure. There are always going to be costs associated with having a longer, vulnerable juvenile period. The evolution of a complex social system in modern humans may have developed, in part, as a means of minimizing this problem and countering factors like high predation pressure, not only from other species, but also from within the same species. It is possible that in the case of neanderthals, during the period when they coexisted with early *Homo sapiens* 100,000–35,000 years ago, factors existed such that the longer juvenile phase was just too long, resulting in unacceptably high levels of juvenile mortality.

Based on this indirect method of interpreting the variable timing of the transition from the infantile to juvenile phases, the dominant trend in hominid evolution can be interpreted as being one of sequential hypermorphosis—the stretching out of all growth phases. The assessment that onset of sexual maturity was also progressively delayed means that terminal hypermorphosis was also operating. This is unlike lamprey evolution (Chap. 10), where there is only delay in transition from one prejuvenile phase to another without concomitant delay in the last transition, from juvenile to adult. In hominids the result is extension of faster embryonic, infantile, and juvenile growth rates in certain structures, principally the skull and brain, the lower limbs and foot bones, as well as body size as a whole. But dissociation of skull development from the jaw, upper limbs, and digits in the lower limbs has resulted in the evolution of some paedomorphic characters. Like so many other species of animals, *Homo sapiens* displays a cocktail of features, some of which have developed more than in the ancestor, others less so. Developmental trade-offs have overwhelmingly favored peramorphic features.

It could, of course, be argued that this entire premise is based an a number of inferences of relationships between brain size and other characteristics that would not hold up to scrutiny, given direct evidence. But as I mentioned earlier, techniques do exist that allow actual ages of fossil mammalian material to be calculated. One that has gained some popularity in recent times, particularly with respect to hominid dental development, is the sectioning of teeth and counting incremental growth lines. Another, nondestructive way is counting enamel perikymata (growth lines present on the surface of well-preserved teeth). Using this method, it has been possible to work out ages of death of three young australopithecines that died just as their first molars were erupting. The analysis indicated death at 3.2–3.3 years, which closely matched predictions based on evaluations made from brain capacity.[12]

Another aspect of the evolution of hominids is their apparent change in diet, from being exclusively vegetarian, in the case of australopithecines, to omnivorous, in species of *Homo*, with the inclusion of a significant proportion of animal products. Even this change may have its roots in the sequential hypermorphosis that has driven hominid evolution. As I have mentioned earlier, growing brain tissue is metabolically a very expensive exercise. Although the brain weighs only about 2 percent of the total body weight, it uses about 17 percent of the body's energy. Hominid evolution is characterized by a more than threefold increase in brain volume in a little over four million years (not a very long period of time, geologically and evolutionarily speaking), and so the energy required to produce proportionately more brain tissue had to be found in different ways. There could have been a simple increase in energy input. This would have required consumption of more of the same type of food—shoveling in more leaves and roots. Or it could have occurred by changing to a more high-quality diet. Another possibility is that there was a developmental trade-off, a feature that is, I believe, so characteristic of much of evolution. In the case of hominids it would mean that there must have been a substantial reduction in certain structures. For such a significant increase in brain size in hominids (compounded by the larger body size), another option is more likely—a combination of both increased energy input and a developmental trade-off.

Leslie Aiello of University College in London, and Peter Wheeler of Liverpool John Moores University in England, have come up with the intriguing suggestion that this developmental trade-off in hominid evolution was between increased brain size and decreased size of the gut. Among the metabolically expensive organs of the body, in addition to the brain, are the so-called *splanchnic organs* (the liver and gastro-intestinal tract). Compared with other primates, we, to put it simply, lack guts. As a function of our body size our gut is very small, being 60 percent of what we should have, in terms of our body size. This might seem to be counterproductive to evolving a larger brain, but as Aiello and Wheeler have pointed out, gut size is highly correlated with diet. On the whole, small guts only function with high-quality, easy-to-digest food (in other words, lean, high-protein animal products), whereas large guts are needed to process large amounts of lower nutritional vegetative matter. In their "expensive-tissue hypothesis," Aiello and Wheeler argue strongly that the rapid evolution in brain size in hominids may lie in what we eat. For we, as large-brained primates, unexpectedly have relatively low metabolic rates.[13]

If selection has so favored our large brains, then one of the developmental trade-offs could have been with the gut. But to function and produce the necessary levels of energy to grow such a large brain, there would have had to have been a major change in the type of food consumed. Wear patterns on the teeth of australopithecines suggest that they subsisted on a diet of leaves and fleshy fruits. However, even the early species of *Homo* are thought to have consumed meat as part of their diet. Evidence for this comes from polish on tools and by cut marks found on bones thought to have come from their meals. *Homo erectus* is generally thought to have been more predatory in its habits than the earliest *Homo* species, and therefore to have eaten more meat.[14]

There is also evidence from changing skeletal shape to support the view that the gut reduced in size during hominid evolution. When reconstructed, the rib cage of *Australopithecus afarensis* looks very chimpanzee-like, in being funnel-shaped, wider at the base than at the top, to incorporate a larger gut. This shape is found in chimpanzees that mainly, although not exclusively, are vegetarians. They both also have a wider pelvis than possessed by later species of *Homo*. *Homo ergaster* was the first hominid to have the narrower pelvis and and more barrel-shaped rib cage, implying a smaller gut and a change in diet. With this change in diet we see the development of the use of tools that, to a large extent, were probably initially used in butchering. Such a manifestation of a more complex behavior, involving tool use and probably complex methods to catch their food, was all part of the evolution of the larger brain. The evolution of our physiology was clearly a highly complex process. The developmental trade-off of a peramorphic brain and a paedomorphic gut typify the overwhelming importance of changes in the timing of development as one of the most, if not the most, significant factors controlling our evolutionary history. And with the evolution of a bigger brain came the evolution of much more complex behaviors.

Oh, What a Tangled Web Our Brains Have Woven

"Although a prolonged period of juvenile helplessness and dependency would, by itself, be disadvantageous to a species because it endangers the young and handicaps their parents, it is a help to man because the slow development provides time for learning and training, which are far more extensive and important in man than in any other animal." So wrote Theodosius Dobzhansky in 1962 (p. 58) in his book *Mankind*

Evolving. As Kathleen Gibson has argued more recently, the sequential delays in transition from one phase of growth to another—from embryo to infant, to juvenile, to adolescent, to adult—have resulted in the evolution of our large, complex brain, with a correspondingly wide array of behaviors. More significantly, perhaps, it has resulted in greater growth of the cerebral cortex—the seat of conscious thought, of memory, intelligence, and speech. Thus, the longer this part of the brain has to develop, the more complex its cognitive capabilities. In particular, the delay in cessation of embryonic growth has been crucial, because it is during this early phase of development in the womb that many of the cortical brain cells are formed. (At birth, the brain has achieved 25 percent of its adult weight. By five years of age it has grown to 90 percent, then to 95 percent at ten years.)[15] Extensions of the succeeding infantile and juvenile phases produce an even more complex brain, because it is during these periods that dendritic growth of the neurons is occurring. Consequently, of all primates we have by far and away the greatest number of interconnecting neurons. This longer growth also allows more time for generation of synapses, myelination, and blood supply growth.[16] Compared with all other primates, living or extinct, our brain is larger and more complex at the end of each growth phase. Myelination and greater development of synapses are important because they are critical in maturing memory, intelligence, and language skills.[17] Myelination tracks along a very set sequence in primates and in mammals overall, but its timing varies between species. Thus, in rhesus monkeys myelination occurs until about 3.5 years and then stops, whereas in humans it persists well into adolescence.[18] The cessation of dendritic growth is delayed more in humans than in any other primates, persisting until about 20 years of age. As Gibson has written, neurologically there is nothing paedomorphic about the adult human brain. In all aspects it has developed hypermorphically well beyond that of all other primates.

Longer childhood occurs in conjunction with this combination of physical facilities in the brain. It is during this period that learning occurs most rapidly. The meaning of these sequential delays in transition from one phase to another has been misinterpreted. It has led to the attitude that we grow more "slowly" than other primates, meaning that we must therefore be paedomorphic. This is a fundamental misunderstanding of what paedomorphosis is. For in true paedomorphosis the descendant adult remains in the ancestral juvenile state. But we humans develop beyond the ancestral adult condition, particularly in brain size and complexity. As a result we go "beyond" other primates (and presumably our ancestors) in our cognitive abilities. Our larger, more complexly interconnected neocortex has the capacity to store greater amounts of information and to undertake more complex mental constructional functions, such as to process and to articulate a complex language, manufacture intricate tools, and participate in complex social development.

It is generally accepted that our cognitive abilities are directly related to the size of our cortex and the number of neural connections, since this allows more complex mental constructions to be undertaken involving the creation and manipulation of words, objects, and ideas. One of the leading proponents of this peramorphic ("overdeveloped") view of human cognitive evolution is Sue Parker of Sonoma State University in California. She has resurrected concepts articulated in the late nineteenth century by the American psychologist James Baldwin. He suggested that the stages of mental development of humans recapitulated the stages of human and primate evolution. However, trying to pursue such concepts in an intellectual environment where the Haeckelian concepts of recapitulation were beginning their slide into oblivion meant that by the 1920s not only had recapitulation been dismissed, but so too had any idea of the notion of psychological evolution, or of mental evolution in any form whatsoever. As Parker has pointed out, attempts in recent years to infer that more recently evolved hominids "overdeveloped" or "recapitulated" behaviors of earlier primates has tended to be treated with a certain degree of scorn and derision.[19] To many anthropologists it is much more acceptable to follow the ideas, such as those espoused by Ashley Montagu, of ascribing our apparent retention of a "playful spirit of the young" as evidence for our paedomorphic origins. But as Parker has stressed, when Baldwin's ideas, heretical as they may seem, are looked at on the basis of the much more detailed analyses of cognitive development carried out in recent years on a range of primates, including humans, and looked at wearing heterochronic eyeglasses, then the evolution of mentality can be analyzed more objectively.

Today, much of our framework of developmental psychology is based on research carried out over 50 years by the Swiss-French psychologist Jean Piaget. Working initially on his own children, and later on children at the Jean-Jacques Rousseau Institute in Geneva, Piaget and his colleagues devised a classification of the stages of infantile and juvenile development. Piaget was a developmental psychologist who attempted to relate the ontogeny of cognitive abilities to the history of science. He recognized that cognitive development is epigenetic, that is to say, each developmental stage was contingent, and built upon, the differentiation and coordination of the preceding stage. Piaget recognized certain ontogenetic cognitive stages in human development. The first period, from birth to 2 years, is called the *sensorimotor period*. This is followed by the *preoperations period* from 2 to 6 years, then the *concrete operations period*, from 6 to 12 years. Lastly comes the *formal operations period*, from 12 years to adulthood. Within each of these periods there is a set sequence of events that crosses a range of cognitive abilities, involving objects, space, time, causality, and imitation. In the post-sensorimotor periods, Piaget also included aspects of classification, number, geometry, and chance.[20] In terms of our appreciation of the importance of sequential hypermorphosis in hominid evolution, the earliest period, the sensorimotor, is perhaps the most illuminating.

In humans this period covers the first two years of life, during which children experience the practical discovery of the world around them: of the properties of objects, space, time, and, to some extent, causality. They discover relationships between objects and learn new concepts through imitation as a means of assimilating new information. The sequence of learning within this period can be divided into six stages in six series or domains: sensorimotor intelligence, space, time, causality, imitation, and object concept. So, for example, from 2 to 4 months of age infants characteristically use repeated actions, like thumb sucking. Between 3 and 8 months they undertake repetitive actions, but ones which by chance produce some kind of reaction, such as "shake a rattle and it makes a noise." In later stages, such as from 12 to 18 months, the repeated actions become more complex. They involve those that are contingent upon variations on a theme of actions, like seeing how far food can be thrown from the highchair. Each of these series can be further subdivided into six-stage sequences. The preoperations period, from 2 to 6 years, is intermediate in terms of sensorimotor and operative intelligence. Here imitative behavior becomes more complex and develops into symbolic play, as well as imitation of drawing, speaking, and mental imagery. In the following concrete operations period, the 6- to 12-year-olds develop, among others, concepts of the conservation of quantity, weight, and volume (my seven-year-old son worked out for himself today, while he was helping me with the washing up, why the level of the water in the sink rose when he put plates into it). Symbolic play becomes a game with rules; and an understanding of unseen mediating forces develops. In the final, formal operations period, among other aspects, hypothetical deductive systems and ways of developing alternative hypotheses are formed. While there has been much debate among developmental psychologists since its inception as to the accuracy of Piaget's scheme, it still proves to be the most effective method for understanding patterns of human cognitive development. Furthermore, it has provided the basis for later developmental studies. Moreover, the detail with which the ontogenetic stages are characterized means that it provides an excellent framework for interpreting variations in the patterns of primate cognitive evolution.

The behaviors of many primates have been well studied, especially the great apes, baboons, and some macaques in particular. These studies have revealed patterns of similarity and dissimilarity, especially with regard to the lengths spent in each developmental stage, as well as the number of stages passed through. In general terms, when other primates are compared with humans it can be seen that they pass through each of the same six behavioral series, but usually they do not develop as far within each. As Parker has pointed out, the adults of great apes display a range of cognitive abilities that are comparable with those found in humans from two to four years old. Adult cebus monkeys develop cognitive abilities similar to those of a two-year-old human child. Even fewer developmental stages are passed through by adult macaques, which show cognitive abilities like those of a one-year-old child. What humans are doing is passing through the same stages of development as other primates, not only faster, but also remaining in each stage for longer and then going "beyond" other primates. Within primates, the descendants of the evolutionarily earliest radiation, the lorises and lemurs, possess reflex grasping and simple manipulative abilities, consistent with the first two stages of sensorimotor development, but they do not pass to the next stages.

Monkeys have greater abilities than these, but less than those of apes. The stump-tailed macaque is able to complete the fifth stage of the object concept series (an object is shown, then hidden), and reaches the fourth stage of the sensorimotor series (things are pulled apart). But it does not pass to the next stage of tool use or imitative behavior. Apes, however, go beyond this period into the symbolic substage of preoperations. However, all great apes use tools, but only chimpanzees and orangs use them, as far as we know, in the wild.

Some adult great apes have a cognitive capacity similar to a two- to three-year-old human. They can use tools and have an understanding of the effect of their use. They can imitate new behaviors; have a basic understanding of symbolism, in drawing and symbolic play; and have an understanding of classification and numbers. Macaques, on the contrary, exhibit none of these features. Obviously, there will also be differences within the great apes in the extent to which these various capabilities have evolved. Some chimpanzees are known to use tools in the wild, whereas gorillas do not.

Unraveling the evolution of such behaviors in our ancestors is clearly a most difficult task. But archeological evidence does provide some indirect evidence to support the notion that the prevailing trend in hominid behavioral evolution has been one of sequential hypermorphosis of cognitive development. Compared with their smaller bodied, and smaller brained, australopithicine ancestors, early members of *Homo*, such as *Homo habilis*, living nearly two million years ago, are known to have manufactured simple stone tools. While they probably used tools in much the same way as chimpanzees do, in extractive foraging, it is conjectured that these early hominids may also have used them to butcher animals. Little is known of how the earliest hominids subsisted. Absence of stone tools has led to the speculation that they were simply gatherers. These small primates were no doubt much more susceptible to predation, especially from large cats, than were later, larger hominids. The relatively longer arms of early hominids, in contrast to their relatively short leg length, indicates that they were probably facultatively bipedal, spending some of their time on the ground, but still with the ability to climb trees. Moreover the young, with their relatively even longer arms, probably clung to their mothers, much in the same way as great ape infants do today.

The major transition morphologically, and no doubt behaviorally, in hominid evolution occurred with the evolution of *Homo* proper around 2 million years ago. These forms were probably fully bipedal. Indirect evidence comes from the distribution of early hominids. Prior to 1.9 million years ago, the only remains are known from Africa. But recent discoveries near the Yangtze Gorge in China of *Homo erectus* remains of this age show that hominids had become capable of moving long distances.[21] One way, you would think, of being able to establish later hominids' progression through Piaget's scheme of cognitive developmental stages would be the study of the evolution of stone tools. Here, seemingly, is a ready-made tool (as it were) to assess the extent of cognitive development. But the degree of complexity of stone tools remained largely static for nearly 2 million years. Although there was a significant increase in complexity at the onset of Middle Paleolithic times, 125,000 years ago, as *Homo erectus* gave way to the early *Homo sapiens*, the great escalation in complexity didn't really occur until Late Paleolithic times, when regionalized tool cultures became established. This coincided, some 30,000–35,000 years ago, with the blossoming of art and the complex cultural patterns of modern *Homo sapiens*.

Parker suggests that if we recognize Piaget's formal operations period, which is the last of the four basic periods of cognitive development, and if at the other end of hominid evolution we can construe the level of development by comparison with great ape development, then the various levels of cognitive development in between can be assigned. So, for example, Parker and Gibson attribute early concrete operational concepts to those *Homo erectus* that manufactured biface Acheullian handaxes. Australopithecine adults, they believe, attained only the early preoperations periods (equivalent to a 2- to 3-year-old human); early *Homo*, such as *Homo habilis*, developed to the late preoperations period (like a 5- to 6-year-old human child); *Homo erectus*, as I have mentioned, to the early concrete operations period (the human equivalent of 6–8 years); early *Homo sapiens* (or as some would have it, *Homo heidelbergensis* and *Homo neanderthalensis*) to the late concrete operations period (10- to 12-year-old human equivalent).[22] The indications from body size, brain weight, timing of tooth eruption, and maturation times all point to the evolution of cognitive abilities in hominids as one of a steady stretching out of mental and intellectual capabilities, as well as physical development. *Homo habilis* used a simple cracked stone to get his food; *Homo erectus*, a slightly more sophisticated, worked rock. Early *Homo sapiens* created more elaborate stone or bone tools. I, as a late *Homo sapiens*, use the complexity of a computer as my tool to indirectly provide me with food.

As with all other organisms, our evolution has come about not by great genetic upheavals but by subtle changes in our developmental program: an extension of a growth phase here, an extension of a growth phase there; a structure growing a little more, another growing a little less. Over the vast expanse of geological time this has compounded to produce highly significant anatomical, life history, and behavioral changes. In just a little over four million years, this has led from a small, chimpanzee-sized primate that sauntered periodically on the ground in small groups, spending life looking for food and trying to avoid being eaten, to the super ape that we have become today. Our evolution has been channeled by myriad external influences: climate, food, predators, our interrelationships with other organisms; and fueled by heterochrony. What has been selected out of the various developmental possibilities is a stretching out of our growth phases. Obviously, underlying these developmental changes is the genetic control of development, in particular the regulation of those hormones that control our growth and our maturity.

Heterochrony, the changing developmental program that can operate from the moment of our conception and affect all stages of growth, has long been the missing link in evolutionary studies. But when fully taken into consideration, our evolutionary history, like that of all other organisms, can be seen to have been orchestrated by three factors. For evolution is not only about genetics and natural selection. Just as crucial are the changes in the timing and rate of development, with the three, genetics, heterochrony, and natural selection, forming an interdependent evolutionary triumvirate.

Epilogue

IT WAS STILL RAINING. BUT IT WAS TWENTY-FIVE YEARS LATER, IN 1994. And the rain-sodden hills were on the other side of the world, in Guizhou Province, southern China. I had set out with my colleagues Zhou Zhiyi, Zhao Yuanlong, Yin Gongzheng, and Yuan Wenwei early in the day from the little town of Duyun to see, at first hand, what, on paper, looked to be a wonderful example of sequential heterochrony in fossil trilobites. Twenty kilometers north of Duyun lay our destination, the village of Balang. The road to the fossil site wound up a steep-sided valley, the lower slopes lined with neat rows of tea bushes, the upper slopes shrouded in omnipresent clouds. The valley became narrower and the road shrank to a single track. Then after shrugging around the last bend, it all but disappeared into a sea of rice fields as the valley opened out. Through the rain we could see at the far side of the valley the thatched huts of Balang. We almost reached it before the axles sank into the mud.

It was hard to know what to expect. The previous year I had been given, by my colleague trilobite paleontologist Zhou Zhiyi at the Nanjing Institute of Geology and Palaeontology, a draft manuscript of a paper by one of his students, Yu Feng, and asked to collaborate. What Yu had found in a sequence of rocks, originally laid down as mud in the same Early Cambrian sea that had also covered northwest Scotland, was a superb evolutionary sequence of oryctocephalid trilobites. These were complete specimens of a number of species appearing successively higher in the layers of rocks, from the tiniest larvae, just hatched from the egg, through juveniles to adults. It was as superb an ontogenetic and evolutionary sequence as you could ever hope to find anywhere.

We made our way out of the village along the pathway that leads into the next valley. By the side of the track, which had been cut into the white camellia-covered hillside, were gently dipping shales in which the trilobites were lurking. All that was necessary was gently to crack the shales with a hammer—then, ever so carefully, pull apart the layers of the rock, as if we were opening a very old and precious book. As the rain continued to fall, perfect, complete images of trilobites were revealed that had last felt water easing over them more than half a billion years before. As we continued to collect along the track, we were traveling up through Early Cambrian time, perhaps a couple of million years in a few hours. It soon became obvious that the evolutionary processes that had acted on these trilobites so long ago were, as I had suspected, much the same as those that had also driven the evolution of the hominids who, more than half a billion years later, were collecting them. As with our own evolution, the development of successively younger species of trilobites had been progressively stretched out.

These trilobites became extinct long ago: our fate, no doubt, sometime in the future. As a paleontologist, it is tempting, sometimes, to look, like the Roman god Janus, not only back into the past but also into the future, into our evolutionary destiny. Why not, for once, peer into the crystal ball and speculate what our descendants might look like four million years from now, if the broad, generalized trends that have marked our evolutionary history continue into the future?

After all, we just happen to be around in this particular instant of geological time. Looking at the evolution of the first four species in the trilobite lineage, it wouldn't have been too difficult to predict what the last species was going to look like. Of course, to undertake such predictive evolutionary analysis presupposes that our unique species characteristic of drastically changing our environment will not adversely affect the internal forces that have driven our evolution to where it has arrived today.

As in humans, sequential hypermorphosis led to the evolution of this Early Cambrian trilobite, *Arithrocephalus balingensis*, from China. Drawing by Yu Feng.

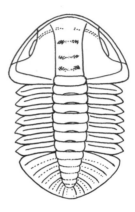

So, who knows? Maybe, if you were able to come back four million years from now, and, if our descendants can manage to find enough high energy brain food (without losing too much gut), you might be met by people with a massive head, carrying a brain perhaps three times larger than ours, along with much-reduced jaw and teeth. (To get enough energy to support this brain, along with having to compensate for the reduced teeth, I wonder whether they might not be reduced to subsisting entirely on a diet of blood!) Standing a good two and a half meters tall, with massive legs, proportionately bigger feet, but tiny toes, these people would remain as children until well into their twenties. And of course, accompanying these changes would be the achievement of one of life's age-old quests—how to live longer. Our *Homo "giganteus"* might be expected to live until he and she were perhaps 150 years old. However, whether the current hominid species will have left enough of a habitable planet to allow such wild speculation ever actually to happen, only time will tell.

Notes

1. The Evolving Embryo

1. S. J. Gould, *Ontogeny and Phylogeny* (Cambridge, Mass.: Harvard University Press, Belknap Press, 1977); Albrecht von Haller, *Hermanni Boerhaave praelectiones academicae*, vol. 5, pt. 2, trans. H. B. Adelmann in *Marcello Malpighi and the Evolution of Embryology*, 5 vols. (Ithaca: Cornell University Press, 1966).

2. J. D. Y. Peel, *Herbert Spencer: the Evolution of a Sociologist* (London: Heinemann, 1971).

3. P. J. Bowler, "The Changing Meaning of 'Evolution,'" *Journal of the History of Ideas* 36 (1975): 95–114; Gould, *Ontogeny and Phylogeny*.

4. Phylogeny can be defined as the evolutionary history of a particular lineage of species.

5. E. Szathmáry and J. Maynard Smith, "The Major Evolutionary Transitions," *Nature* 374 (1995): 227–32.

6. D. W. McShea, "Evolutionary Change in the Morphological Complexity of the Mammalian Vertebral Column," *Evolution* 47 (1993): 730–40.

7. G. Boyajian and T. Lutz, "Evolution of Biological Complexity and Its Relation to Taxonomic Longevity in the Ammonoidea," *Geology* 20 (1992): 983–86.

2. The Topsy-Turvy World of Dr. Haeckel and Dr. Garstang

1. H. F. Osborn, *From the Greeks to Darwin* (New York: Scribner, 1929).

2. S. J. Gould, *Ontogeny and Phylogeny* (Cambridge, Mass.: Harvard University Press, Belknap Press, 1977).

3. Lorenz Oken, *Lehrbuch der Naturphilosophie*, 3 vols. (Jena: F. Frommand, 1809–11).

4. M. L. McKinney and K. J. McNamara, *Heterochrony: The Evolution of Ontogeny* (New York: Plenum Press, 1991).

5. D. T. Donovan, "The Influence of Theoretical Ideas on Ammonite Classification from Hyatt to Trueman," *University of Kansas Paleontological Contributions* 62 (1973): 1–16.

6. W. Garstang, "The Morphology of the Tunicata, and Its Bearing on the Phylogeny of the Chordata," *Quarterly Journal of Microscopical Science* 75 (1928): 51–187; W. Garstang, "The Theory of Recapitulation: A Critical Restatement of the Biogenetic Law," *Journal of the Linnaean Society* 35 (1922): 81–101.

7. Many biologists and paleontologists, even to the present day, use the terms *neoteny* and *paedomorphosis* interchangeably. But as Stephen Gould first showed in 1977, the two terms mean quite different things. Neoteny describes a process of reducing a growth rate. The *effect* of neoteny is to produce paedomorphosis.

8. *Ammocoetes* are the long-lived, blind, toothless larvae of the "primitive" vertebrates, the lampreys.

9. *Perennibranchiate:* retaining gills throughout juvenile and adult life.

10. A *notochord* is the elastic, cellular axial support formed beneath the nerve chord in the early embryo of all vertebrates, later surrounded or replaced by the vertebrae in most vertebrates.

11. P. Alberch, S. J. Gould, G. F. Oster, and D. B. Wake, "Size and Shape in Ontogeny and Phylogeny," *Paleobiology* 5 (1979): 296–317.

3. The Shape of Things to Come

1. L. Wolpert, "Pattern Formation in Biological Development," *Scientific American* 239, no. 4 (1978): 124–37; L. Wolpert, "Pattern Formation and Change," in J. T. Bonner, ed., *Evolution and Development* (Berlin: Springer-Verlag, 1982); L. Wolpert, *The Triumph of the Embryo* (New York: Oxford University Press, 1991).

2. S. B. Carroll, "Homeotic Genes and the Evolution of Arthropods and Chordates," *Nature* 376 (1995): 479–85.

3. K. J. McNamara, "The Role of Heterochrony in the Evolution of Cambrian Trilobites," *Biological Reviews* 61 (1986): 121–56.

4. N. P. Patel, B. G. Condron, and K. Zinn, "Pair-Rule Expression Patterns of *Even-skipped* Are Found in Both Short- and Long-germ Beetles," *Nature* 367 (1994): 429–34.

5. L. Wolpert, "The Shape of Things to Come," *New Scientist*, 27 June 1992, 38–42.

6. B. K. Hall, *Evolutionary Developmental Biology* (London: Chapman and Hall, 1992).

7. J. J. Henry and R. M. Grainger, "Early Tissue Interactions Leading to Embryonic Lens Formation in *Xenopus laevis*," *Developmental Biology* 141 (1990): 149–63.

8. J. A. Long, *The Rise of Fishes: 500 Million Years of Evolution* (Baltimore: Johns Hopkins University Press, 1995).

9. M. M. Smith, "Heterochrony in the Evolution of Enamel in Vertebrates," in K. J. McNamara, ed., *Evolutionary Change and Heterochrony* (Chichester and New York: Wiley, 1995), 125–50.

10. J. H. Marden and M. G. Kramer, "Surface-skimming Stoneflies: A Possible Intermediate Stage in Insect Flight Evolution," *Science* 266 (1994): 427–30.

11. S. B. Carroll, S. D. Weatherbee, and J. A. Langeland, "Homeotic Genes and the Regulation and Evolution of Insect Wing Number," *Nature* 375 (1995): 58–61.

12. K. Basler and G. Struhl, "Compartment Boundaries and the Control of *Drosophila* Limb Pattern by *hedgehog* Protein," *Nature* 368 (1994): 208–14.

13. B. Shea, R. E. Hammer, R. L. Brinster, and M. J. Ravosa, "Relative Growth of the Skull and Postcranium in Giant Transgenic Mice," *Genetics Research* 56 (1990): 21–34.

14. A. Ishikawa and T. Namikawa, "Postnatal Growth and Development in Laboratory Strains of Large and Small Musk Shrews (*Suncusmurinus*)," *Journal of Mammalogy* 68 (1987): 766–74.

15. J. F. Fallon and J. Cameron, "Interdigital Cell Death during Limb Development of the Turtle with an Interpretation of Evolutionary Significance," *Journal of Embryological Experimental Morphology* 40 (1977): 285–89.

4. It's a Dog's Life

1. S. J. Gould, "A Biological Homage to Mickey Mouse," *Natural History* 88, no. 5 (1979): 30–36.

2. D. I. Perrett, K. A. May, and S. Yoshikawa, "Facial Shape and Judgements of Female Attractiveness," *Nature* 368 (1994): 239–42.

3. R. K. Wayne, "Cranial Morphology of Domestic and Wild Canids: The Influence of Development on Morphological Change," *Evolution* 40 (1986): 243–61.

4. B. T. Shea, "Dynamic Morphology: Growth, Life History, and Ecology in Primate Evolution," in C. J. DeRousseau, ed., *Primate Life History and Evolution* (New York: Wiley-Liss, 1990), 325–52.

5. G. K. Creighton and R. E. Strauss, "Comparative Patterns of Growth and Development in Cricetine Rodents and the Evolution of Ontogeny," *Evolution* 40 (1986): 94–106.

6. D. C. Smith, "Heritable Divergence of *Rhagoletis pomonella* Host Races by Seasonal Asynchrony," *Nature* 336 (1988): 66–67; J. L. Feder, C. A. Chilcore, and G. L. Bush, "Genetic Differentiation between Sympatric Host Races of the Apple Maggot Fly, *Rhagoletis pomonella*," *Nature* 336 (1988): 61–64.

7. E. A. Bernays, "Diet-induced Head Allometry among Foliage-chewing Insects and Its Importance for Graminovores," *Science* 231 (1986): 495–97.

8. R. F. Leclerc and J. C. Regier, "Heterochrony in Insect Development and Evolution," *Developmental Biology* 1 (1990): 271–79.

9. M. Ashburner, "Chromosomal Action of Ecdysone," *Nature* 285 (1980): 435–36.

10. D. E. Wheeler and H. F. Nijhout, "Soldier Determination in Ants: New Role for Juvenile Hormone," *Science* 213 (1981): 361–63; Y. Roisin, "Morphology, Development, and Evolutionary Significance of the Working Stages in the Caste System of *Prorhinotermes* (Insecta, Isoptera)," *Zoomorphology* 107 (1988): 339–47.

11. R. N. Harris, "Density-dependent Paedomorphosis in the Salamander *Notophthalmus viridescens dorsalis*," *Ecology* 68 (1987): 705–12.

12. J. P. Collins and J. E. Cheek, "Effect of Food and Density on Develop-

ment of Typical and Cannibalistic Salamander Larvae in *Ambystoma tigrinum nebulosum*," *American Zoology* 23 (1983): 77–84.

13. P. Alberch, "Possible Dogs," *Natural History* 95, no. 12 (1986): 4–8.

5. Time for Sex

1. J. Hunter, "Account of an Extraordinary Pheasant," *Philosophical Transactions of the Royal Society* 70 (1780): 527–35; I. P. F. Owens and R. V. Short, "Hormonal Basis of Sexual Dimorphism in Birds: Implications for New Theories of Sexual Selection," *Trends in Ecology and Evolution* 10 (1995): 44–47.

2. R. Holmes, "Still Life in Mouldy Bread," *New Scientist*, 26 March 1994, 39–41.

3. J. W. Schopf, "Microfossils of the Early Archean Apex Chert: New Evidence of the Antiquity of Life," *Science* 260 (1993): 640–46.

4. D. Sagan and L. Margulis, "Bacterial Bedfellows," *Natural History* 3 (1987): 26–33.

5. T.-M. Han and B. Runnegar, "Megascopic Eukaryotic Algae from the 2.1-billion-year-old Negaunee Iron-Formation, Michigan," *Science* 257 (1992): 232–35.

6. L. Margulis and D. Sagan, *Origins of Sex: Three Billion Years of Genetic Recombination* (New Haven: Yale University Press, 1986).

7. Meiosis is the cellular mechanism whereby certain cells lose half their complement of chromosomes and become gametes, either eggs or sperm.

8. L. R. Cleveland, "The Origin and Evolution of Meiosis," *Science* 105 (1947): 287–88. Centromeres are the point at which the paired chromatids, which comprise the chromosome, are joined, somewhere along their length.

9. D. Joly, C. Bressac, and D. Lachaise, "Disentangling Giant Sperm," *Nature* 377 (1995): 202.

10. U. Mittwoch, A. M. C. Burgess, and P. J. Baker, "Male Sexual Development in 'a Sea of Oestrogen,'" *The Lancet* 342 (1993): 123–24.

11. U. Mittwoch, "Blastocysts Prepare for the Race to Be Male," *Human Reproduction* 8 (1993): 1550–55.

12. P. Jarman, "Mating System and Sexual Dimorphism in Large, Terrestrial, Mammalian Herbivores," *Biological Reviews* 58 (1983): 485–520. Bovids are a group of artiodactyls that include cattle, bison, ox, sheep, goats, and antelopes. They have true horns (usually in both sexes) that are a simple unbranched core of bone with a sheath of horn; no part is ever shed.

13. B. T. Shea, "Allometry and Heterochrony in the African Apes," *American Journal of Physical Anthropology* 62 (1983): 275–89.

14. R. Z. German, D. W. Hertweck, J. E. Sirianni, and D. R. Swindler, "Heterochrony and Sexual Dimorphism in the Pigtailed Macaque (*Macaca nemestrina*)," *American Journal of Physical Anthropology* 93 (1994): 373–80.

15. J. M. Cheverud, P. Wilson, and W. P. J. Dittus, "Primate Population Studies at Polonnaruwa: III. Somatometric Growth in a Natural Population of Toque Macaques (*Macaca sinica*)," *Journal of Human Evolution* 23 (1992): 51–77.

16. S. R. Leigh, "Patterns of Variation in the Ontogeny of Primate Body Size Dimorphism," *Journal of Human Evolution* 23 (1992): 27–50.

17. J. Lützen, "Unisexuality in the Parasitic Family Entoconchidae (Gastropoda: Prosobranchia)," *Malacologia* 7 (1968): 7–15.

18. P. J. Gullan and A. Cockburn, "Sexual Dichroism and Intersexual Phoresy in Gall-forming Coccoids," *Oecologia* (Berlin) 68 (1986): 632–34.

19. A. Meyer, J. M. Morrissey, and M. Schartl, "Recurrent Origin of a Sexually Selected Trait in *Xiphophorus* Fishes Inferred from Molecular Phylogeny," *Nature* 368 (1994): 539–42.

20. J. B. Hutchins, "Sexual Dimorphism in the Osteology and Myology of Monacanthid Fishes," *Records of the Western Australian Museum* 15 (1992): 739–47.

21. F. Vollrath and G. A. Parker, "Sexual Dimorphism and Distorted Sex Ratios," *Nature* 360 (1992): 156–59.

22. D. Cook, "Sexual Selection in Dung Beetles: I. A Multivariate Study of the Morphological Variation in Two Species of *Onthophagus* (Scarabaeidae: Onthophagini)," *Australian Journal of Zoology* 35 (1987): 123–32.

23. L. Brown and L. L. Rockwood, "On the Dilemma of Horns," *Natural History* 95, no. 7 (1986): 54–61.

24. K. J. McNamara, "Sexual Dimorphism: The Role of Heterochrony," in K. J. McNamara, ed., *Evolutionary Change and Heterochrony* (Chichester and New York: Wiley, 1995).

6. Birds, Brachiopods, and Bushbucks

1. A. Moorehead, *Darwin and the Beagle* (London: Hamish Hamilton, 1969).

2. H. L. Gibbs and P. R. Grant, "Oscillating Selection on Darwin's Finches," *Nature* 327 (1987): 511–13.

3. P. T. Boag, "The Heritability of External Morphology in Darwin's Ground Finches (*Geospiza*) on Isla Daphne Major, Galápagos," *Evolution* 37 (1983): 877–94.

4. Before you get any similar ideas, fossils are protected under Western Australian State legislation and cannot be collected without a permit.

5. K. J. McNamara, "The Earliest *Tegulorhynchia* (Brachiopoda: Rhynchonellida) and Its Evolutionary Significance," *Journal of Paleontology* 57 (1983): 461–73.

6. J. Kingdon, *East African Mammals—An Atlas of Evolution in Africa*, vol. 3, pt. C (*Bovids*) (Chicago: University of Chicago Press, 1982).

7. The Peter Pan Syndrome

1. G. de Beer, *Embryos and Ancestors* (Oxford: Clarendon, 1958).

2. K. J. McNamara, "The Abundance of Heterochrony in the Fossil Record," in M. L. McKinney, ed., *Heterochrony in Evolution: A Multidisciplinary Approach* (New York: Plenum, 1988), 287–325.

3. D. J. Varricchio, "Bone Microstructure of the Upper Cretaceous Theropod *Troodon formosus*," *Journal of Vertebrate Paleontology* 13 (1993): 99–104.

4. T. Cavalier-Smith, "Nuclear Volume Control by Nucleoskeletal DNA, Selection for Cell Volume and Cell Growth Rate, and the Solution of the DNA C-value Paradox," *Journal of Cell Science* 34 (1978): 247–78.

5. S. K. Sessions and A. Larson, "Developmental Correlates of Genome Size in Plethodontid Salamanders and Their Implications for Genome Evolution," *Evolution* 41 (1987): 1239–51.

6. A. Morescalchi and V. Serra, "DNA Renaturation Kinetics in Some Paedogenetic Urodeles," *Experientia* 30 (1974): 487–89.

7. H. A. Horner and H. C. Macgregor, "C-value and Cell-volume: Their Significance in the Evolution and Development of Amphibians," *Journal of Cell Science* 63 (1983): 135–46.

8. J. A. Long, *The Rise of Fishes: 500 Million Years of Evolution* (Baltimore: Johns Hopkins University Press, 1995).

9. W. E. Bemis, "Paedomorphosis and the Evolution of the Dipnoi," *Paleobiology* 10 (1984): 293–307.

10. K. S. Thomson, "Estimation of Cell Size and DNA Content in Fossil Fishes and Amphibians," *Journal of Experimental Zoology* 205 (1972): 315–20.

11. R. W. Williams, C. Cavada, and F. Reinoso-Suárez, "Rapid Evolution of the Visual System: A Cellular Assay of the Retina and Dorsal Lateral Geniculate Nucleus of the Spanish Wildcat and Domestic Cat," *Journal of Neuroscience* 13 (1993): 208–28.

12. S. Conway Morris and J. S. Peel, "Articulated Halkieriids from the

Lower Cambrian of North Greenland and Their Role in Early Protostome Evolution," *Philosophical Transactions of the Royal Society of London* B 347 (1995): 305–58.

13. K. J. McNamara and N. H. Trewin, "A Euthycarcinoid Arthropod from the Silurian of Western Australia," *Palaeontology* 36 (1993): 319–35.

14. B. Swedmark, "The Interstitial Fauna of Marine Sand," *Biological Reviews* 39 (1964): 1–42.

15. E. A. Mancini, "Origin of Micromorph Faunas in the Geologic Record," *Journal of Paleontology* 52 (1978): 311–22.

16. F. Surlyk, "Morphological Adaptations and Population Structures of the Danish Chalk Brachiopods (Maastrichtian, Upper Cretaceous)," *Biologiske Skrifter, Det Kongelige Danske Videnskabernes Selskab* 19 (1972): 1–57.

17. D. Korn, "Impact of Environmental Perturbations on Heterochronic Development in Palaeozoic Ammonoids," in K. J. McNamara, ed., *Evolutionary Change and Heterochrony* (Chichester and New York: Wiley, 1995), 245–60.

18. F. G. Howarth, "High-stress Subterranean Habitats and Evolutionary Change in Cave-inhabiting Arthropods," *American Naturalist* 142, Suppl. (1993): S65–S77.

19. S. Conway Morris and D. W. T. Crompton, "The Origins and Evolution of the Acanthocephala," *Biological Reviews* 57 (1982): 85–115.

8. Images of the Past, Shapes of the Future

1. G. de Beer, "The Evolution of Ratites," *Bulletin of the British Museum (Natural History), Zoology* 4 (1956): 59–70.

2. B. C. Livezey, "Heterochrony and the Evolution of Avian Flightlessness," in K. J. McNamara, ed., *Evolutionary Change and Heterochrony* (Chichester and New York: Wiley, 1995), 169–93.

3. B. C. Livezey, "Flightlessness in the Galápagos Cormorant (*Compsohalieus* [*Nannopterum*] *harrisi*): Heterochrony, Giantism, and Specialization," *Zoological Journal of the Linnean Society* 105 (1992): 155–224.

4. H. E. Strickland and A. G. Melville, *The Dodo and Its Kindred; or, The History, Affinities, and Osteology of the Dodo, Solitaire, and Other Extinct Birds of the Islands Mauritius, Rodriguez, and Bourbon* (London: Reeve, Banham and Reeve, 1848).

5. R. A. Thulborn, "Birds as Neotenous Dinosaurs," *Records of the New Zealand Geological Survey* 9 (1985): 90–92.

6. M. A. Norell, J. M. Clark, D. Demberelyin, B. Rhinchen, L. M. Chiappe, A. R. Davidson, M. C. McKenna, P. Altangerel, and M. J. Novacek,

"A Theropod Dinosaur Embryo and the Affinities of the Flaming Cliffs Dinosaur Eggs," *Science* 266 (1994): 779–82.

7. L. D. Martin, "Mesozoic Birds and the Origin of Birds," in H.-P. Schultze and L. Trueb, eds., *Origins of the Higher Groups of Tetrapods* (Ithaca: Cornell University Press, 1991), 485–540.

8. A. J. Jeram, P. A. Selden, and D. Edwards, "Land Animals in the Silurian: Arachnids and Myriapods from Shropshire, England," *Science* 250 (1990): 658–61.

9. M. M. Coates and J. A. Clack, "Polydactyly in the Earliest Known Tetrapod Limbs," *Nature* 347 (1990): 66–69.

10. J. A. Long, *The Rise of Fishes: 500 Million Years of Evolution* (Baltimore: Johns Hopkins University Press, 1995).

11. N. Shubin and P. Alberch, "A Morphogenetic Approach to the Origin and Basic Organization of the Tetrapod Limb," *Evolutionary Biology* 20 (1986): 319–87.

12. P. Sordino, F. van der Hoeven, and D. Duboule, "*Hox* Gene Expression in Teleost Fins and the Origin of Vertebrate Digits," *Nature* 375 (1995): 678–81.

13. M. Lee, "The Turtle's Long-Lost Relatives," *Natural History* 103, no. 6 (1994): 63–65.

14. K. J. McNamara, "Paedomorphosis in Middle Cambrian Xystridurine Trilobites from Northern Australia," *Alcheringa* 5 (1981): 209–24.

9. Evolving the Shapes Beyond

1. A. Hallam, "Evolutionary Size Increase and Longevity in Jurassic Bivalves and Ammonites," *Nature* 258 (1975): 493–97.

2. R. D. Stevenson, M. F. Hill, and P. J. Bryant, "Organ and Cell Allometry in Hawaiian *Drosophila:* How to Make a Big Fly," *Proceedings of the Royal Society of London* B259 (1995): 105–10.

3. R. A. Coria and L. Salgado, "A New Giant Carnivorous Dinosaur from the Cretaceous of Patagonia," *Nature* 377 (1995): 224–26.

4. J. A. Long and K. J. McNamara, "Heterochrony in Dinosaur Evolution," in K. J. McNamara, ed., *Evolutionary Change and Heterochrony* (Chichester and New York: Wiley, 1995), 151–68.

5. D. J. Varricchio, "Bone Microstructure of the Upper Cretaceous Theropod *Troodon formosus*," *Journal of Vertebrate Paleontology* 13 (1993): 99–104.

6. R. T. Bakker, M. Williams, and P. Currie, "*Nanotyrannus*, a New Genus

of Pygmy Tyrannosaur, from the Latest Cretaceous of Montana," *Hunteria* 1, no. 5 (1988): 1–30.

7. J. R. Horner and P. J. Currie, "Embryonic and Neonatal Morphology and Ontogeny of a New Species of *Hypacrosaurus* (Ornithischia, Lambeosauridae) from Montana and Alberta," in K. Carpenter, K. F. Hirsch, and J. Horner, eds., *Dinosaur Eggs and Babies* (Cambridge: Cambridge University Press, 1994), 312–36.

8. B. J. MacFadden, "Fossil Horses from '*Eohippus*' (*Hyracotherium*) to *Equus*: Scaling, Cope's Law, and the Evolution of Body Size," *Paleobiology* 12 (1986): 355–69.

9. B. T. Shea, "Allometry and Heterochrony in the African Apes," *American Journal of Physical Anthropology* 62 (1983): 275–89.

10. J. L. Gittleman, "Carnivore Life History Patterns: Allometric, Phylogenetic, and Ecological Associations," *American Naturalist* 127 (1986): 744–71.

11. R. A. Martin, "Energy, Ecology, and Cotton Rat Evolution," *Paleobiology* 12 (1986): 370–82.

12. J. B. Graham, R. Dudley, N. M. Aguilar, and C. Gans, "Implications of the Late Palaeozoic Oxygen Pulse for Physiology and Evolution," *Nature* 375 (1995): 117–20.

13. A. M. Lister, "The Evolution of the Giant Deer, *Megaloceros giganteus* (Blumenbach)," *Zoological Journal of the Linnean Society* 112 (1994): 65–100.

14. M. L. McKinney and R. M. Schoch, "Titanothere Allometry, Heterochrony, and Biomechanics: Revisiting an Evolutionary Classic," *Evolution* 39 (1985): 1352–63.

15. K. Tschanz, "Allometry and Heterochrony in the Growth of the Neck of Triassic Prolacertiform Reptiles," *Palaeontology* 31 (1988): 997–1011.

16. E. O. Guerrant, "Heterochrony in Plants: The Intersection of Evolution, Ecology, and Ontogeny," in M. L. McKinney, ed., *Heterochrony in Evolution: A Multidisciplinary Approach* (New York: Plenum, 1988), 111–33.

17. R. B. Primack, "Relationships among Flowers, Fruits, and Seeds," *Annual Review of Ecology and Systematics* 18 (1987): 409–30.

18. R. A. Adams and S. C. Pedersen, "Wings on Their Fingers," *Natural History* 103, no. 1 (1994): 49–55.

19. R. L. Carroll, *Vertebrate Paleontology and Evolution* (New York: W. H. Freeman, 1988).

20. M. B. Fenton, D. Audet, M. K. Obrist, and J. Rydell, "Signal Strength, Timing, and Self-Deafening: The Evolution of Echolocation in Bats," *Paleobiology* 21 (1995): 229–42.

21. S. C. Bennett, "A Statistical Study of *Rhamphorhynchus* from the Solnhofen Limestone of Germany: Year-Classes of a Single Large Species," *Journal of Paleontology* 69 (1995): 569–80; S. C. Bennett, "The Ontogeny of *Pteranodon* and Other Pterosaurs," *Paleobiology* 19 (1993): 92–106.

10. Evolving a Way of Life

1. K.-Y. Wei, "Allometric Heterochrony in the Pliocene-Pleistocene Planktic Foraminiferal Clade *Globoconella*," *Paleobiology* 20 (1994): 66–84.

2. M. Caron and P. Homewood, "Evolution of Early Planktic Formanifers," *Marine Micropalaeontology* 7 (1983): 453–62.

3. J. R. Bryan, "Life History and Development of Oligocene Larger Benthic Foraminifera: A Test of the Environmental Control on Heterochrony," *Tulane Studies in Geology and Paleontology* 27 (1995): 101–18.

4. M. L. McKinney and J. L. Gittleman, "Ontogeny and Phylogeny: Tinkering with Covariation in Life History, Morphology, and Behaviour," in K. J. McNamara, ed., *Evolutionary Change and Heterochrony* (Chichester and New York: Wiley, 1995), 21–47.

5. M. L. McKinney, "Allometry and Heterochrony in an Eocene Echinoid Lineage: Morphological Change as a Byproduct of Size Selection," *Paleobiology* 10 (1984): 407–19.

6. G. Wray, "Causes and Consequences of Heterochrony in Early Echinoderm Development," in K. J. McNamara, ed., *Evolutionary Change and Heterochrony* (Chichester and New York: Wiley, 1995), 197–223.

7. K. J. McNamara, "Palaeodiversity of Cenozoic Marsupiate Echinoids as a Palaeoenvironmental Indicator," *Lethaia* 27 (1994): 257–68.

8. Anon., *Receipts and Relishes, being a Vade Mecum for the Epicure in the British Isles* (London: Whitbread and Co., 1950).

9. J. C. Hafner and M. S. Hafner, "Heterochrony in Rodents," in M. L. McKinney, ed., *Heterochrony in Evolution: a Multidisciplinary Approach* (New York: Plenum, 1988), 217–35.

10. D. F. Morey, "The Early Evolution of the Domestic Dog," *American Scientist* 82 (1994): 336–47.

11. J. P. Scott, "Critical Periods for the Development of Social Behavior in Dogs," in J. P. Scott, ed., *Critical Periods* (Stroudsburg, Pa.: Dowden, Hutchinson, and Ross, 1978).

12. D. K. Belyaev, "Destabilizing Selection as a Factor in Domestication," *Journal of Heredity* 70 (1979): 301–8.

13. B. Cole, "Size and Behavior in Ants: Constraints on Complexity," *Proceedings of the National Academy of Sciences* 82 (1985): 8548–51.

14. H. Friedmann, *The Cowbirds: A Study in the Biology of Social Parasitism* (Springfield, Ill.: Charles C. Thomas, 1929).

15. F. H. Herrick, "Life and Behaviour of the Cuckoo," *Journal of Experimental Zoology* 1 (1910): 171–233.

16. I. G. Jamieson, "Behavioral Heterochrony and the Evolution of Birds' Helping at the Nest: An Unselected Consequence of Communal Breeding?" *American Naturalist* 133 (1989): 394–406.

17. I. Rowley, *Bird Life* (Sydney: Collins, 1975).

18. H. A. Ford, *Ecology of Birds—an Australian Perspective* (Chipping Norton, NSW: Surrey Beatty and Sons, 1989).

19. M. K. Tarburton and E. O. Minot, "A Novel Strategy of Incubation in Birds," *Animal Behaviour* 35 (1987): 1898–99.

20. R. E. Irwin, "The Evolutionary Importance of Behavioural Development: The Ontogeny and Phylogeny of Bird Song," *Animal Behaviour* 36 (1988): 814–24.

21. D. Maurer and C. Maurer, *The World of the Newborn* (New York: Basic Books, 1988).

11. Fueling the Biological Arms Race

1. K. J. McNamara, "The Significance of Gastropod Predation to Patterns of Evolution and Extinction in Australian Tertiary Echinoids," in B. David, A. Guille, J. P. Firal, and M. Roux, eds., *Echinoderms through Time (Echinoderms Dijon)* (Rotterdam: Balkema, 1994), 785–93.

2. D. Jablonski and D. J. Bottjer, "Onshore-Offshore Evolutionary Patterns in Post-Palaeozoic Echinoderms: A Preliminary Analysis," in R. D. Burke, P. V. Mladenov, P. Lambert, and R. L. Parsley, eds., *Echinoderm Biology* (Rotterdam: Balkema, 1988), 81–90.

3. M. A. Bell, "Stickleback Fishes: Bridging the Gap between Population Biology and Paleobiology," *Trends in Ecology and Evolution* 3 (1988): 320–25.

4. E. E. Werner and D. J. Hall, "Ontogenetic Habitat Shifts in Bluegill: The Foraging Rate–Predation Risk Trade-Off," *Ecology* 69 (1988): 1352–66.

5. B. W. Sweeney and R. L. Vannote, "Population Synchrony in Mayflies: A Predator Satiation Hypothesis," *Evolution* 36 (1982): 810–21.

6. A. Sih, "Predators and Prey Lifestyles: An Evolutionary and Ecological Overview," in W. C. Kerfoot and A. Sih, eds., *Predation—Direct and Indirect Impacts on Aquatic Communities* (Hanover, N.H.: University Press of New England, 1987).

7. J. G. Wilson, "Resource Partitioning and Predation as a Limit to Size

in *Nucula turgida* (Leckenby & Marshall)," *Functional Ecology* 2 (1988): 63–66.

8. M. T. Edley and R. Law, "Evolution of Life Histories and Yields in Experimental Populations of *Daphnia magna*," *Biological Journal of the Linnaean Society* 34 (1988): 309–26.

9. T. A. Crowl and A. P. Covich, "Predator-induced Life-History Shifts in a Freshwater Snail," *Science* 247 (1990): 949–51.

10. K. J. McNamara, "Progenesis in Trilobites," in D. E. G. Briggs and P. D. Lane, eds., *Trilobites and Other Early Arthropods: Papers in Honour of Professor H. B. Whittington, F.R.S.*, Special Papers in Palaeontology, no. 30 (1983): 59–68.

11. S. Conway Morris, "Late Precambrian and Cambrian Soft-bodied Faunas," *Annual Review of Earth and Planetary Sciences* 18 (1990): 101–22.

12. The Baby-faced Super Ape

1. S. J. Gould, *Ontogeny and Phylogeny* (Cambridge, Mass.: Harvard University Press, Belknap Press, 1977).

2. E. D. Cope, *The Primary Factors of Organic Evolution* (Chicago: Open Court, 1896).

3. L. Bolk, *Das Problem der Menschwerdung* (Jena: Gustav Fischer, 1926).

4. A. Montagu, *Growing Young* (New York: McGraw-Hill, 1981).

5. B. T. Shea, "Heterochrony in Human Evolution: The Case for Human Neoteny," *Yearbook of Physical Anthropology* 32 (1989): 69–101.

6. T. G. Bromage, "Ontogeny and Phylogeny of the Human Face," in J. Derousseau, ed., *Primate Life History and Evolution*, Wenner-Gren Foundation, Conference no. 104 (New York: Alan R. Liss, 1989).

7. T. D. White, G. Suwa, and B. Asfaw, "*Australopithecus ramidus*, a New Species of Early Hominid from Aramis, Ethiopia," *Nature* 371 (1994): 306–12.

8. H. M. McHenry, "Body Size and Proportions in Early Hominids," *American Journal of Physical Anthropology* 87 (1992): 407–31.

9. B. H. Smith, "Dental Development and the Evolution of Life History in Hominidae," *American Journal of Physical Anthropology* 86 (1991): 157–74.

10. S. Hartwig-Scherer, "Body Weight Prediction in Early Fossil Hominids: Towards a Taxon-'independent' Approach," *American Journal of Physical Anthropology* 92 (1993): 17–36.

11. S. T. Parker, "Why Big Brains Are So Rare: Energy Costs of Intelligence and Brain Size in Anthropoid Primates," in S. T. Parker and K. R.

Gibson, eds., *"Language" and Intelligence in Monkeys and Apes* (Cambridge: Cambridge University Press, 1990).

12. T. G. Bromage and M. C. Dean, "Re-evaluation of the Age at Death of Immature Fossil Hominids," *Nature* 317 (1985): 525–27.

13. L. C. Aiello and P. Wheeler, "The Expensive-Tissue Hypothesis," *Current Anthropology* 36 (1995): 199–221.

14. P. Shipman and A. Walker, "The Costs of Becoming a Predator," *Journal of Human Evolution* 18 (1989): 373–92.

15. J. M. Tanner, "Human Growth and Development," in S. Jones, R. Martin, and D. Pilbeam, eds., *The Cambridge Encyclopedia of Human Evolution* (Cambridge: Cambridge University Press, 1992).

16. A *synapse* is the junction between two neurons, where one neuron transmits an impulse to the dendrite of another neuron. *Myelination* is the laying down of a fatty myelin sheath around the axon (the long thread that runs out from the cell body and that conducts impulses). Myelinated neurons transmit information up to twenty times faster than nonmyelinated neurons. Thus, a nerve impulse can travel down your spinal cord to the tip of your toe in under 25 milliseconds.

17. K. R. Gibson, "New Perspectives on Instincts and Intelligence: Brain Size and the Emergence of Hierarchical Mental Construction Skills," in S. T. Parker and K. R. Gibson, eds., *"Language" and Intelligence in Monkeys and Apes* (Cambridge: Cambridge University Press, 1990).

18. K. R. Gibson, "Myelination and Behavioral Development: A Comparative Perspective on Questions of Neoteny, Altriciality, and Intelligence," in K. R. Gibson and A. C. Petersen, eds., *Brain Maturation and Cognitive Development* (New York: De Gruyter, 1991).

19. S. T. Parker, "Using Cladisitic Analysis of Comparative Data to Reconstruct the Evolution of Cognitive Development in Hominids," paper presented at the Animal Behavior Society Meetings Symposium on Phylogenetic Comparative Methods, Seattle, Wash., July 1994.

20. J. Piaget, *Play, Dreams, and Imitation in Childhood* (New York: Norton, 1962).

21. W. Huang, R. Ciochon, Y. Gu, R. Larick, Q. Fang, H. Schwarcz, C. Yonge, J. de Vos, and W. Rink, "Early *Homo* and Associated Artefacts from Asia," *Nature* 378 (1995): 275–78.

22. S. T. Parker and K. R. Gibson, "A Developmental Model for the Evolution of Language and Intelligence in Early Hominids," *Behavioral and Brain Science* 2 (1979): 367–408.

Figure Credits

The following figures in this volume are reproduced by permission of the individuals and organizations listed below.

p. 3: reproduced from K. J. McNamara, "Paedomorphosis in Scottish Olenellid Trilobites (Early Cambrian)," *Palaeontology* 21 (1978): 635–55, by permission of the Palaeontological Association.

p. 51 (top and bottom): redrawn by permission from *Nature* 376 (1995): 479–85, S. B. Carroll, "Homeotic Genes and the Evolution of Arthropods and Chordates," © 1995, Macmillan Magazines Ltd.

p. 77: Snoopy © 1958 United Feature Syndicate, Inc.

p. 78: "The Evolution of Mickey Mouse," © Disney Enterprises, Inc.

p. 81: redrawn by permission from R. K. Wayne, "Cranial Morphology of

Domestic and Wild Canids: The Influence of Development on Morphological Change," *Evolution* 40 (1986): 243–61.

p. 114: redrawn by permission from M. L. McKinney and K. J. McNamara, *Heterochrony: The Evolution of Ontogeny* (New York: Plenum, 1991). © 1991, Plenum Publishing Corp.

p. 115: redrawn by permission of Blackwell Science Pty. Ltd. from Andrew Cockburn, *An Introduction to Evolutionary Biology* (Oxford: Blackwell Scientific Publications, 1992). © 1992, Blackwell Science Pty. Ltd.

p. 116: redrawn by permission from M. L. McKinney and K. J. McNamara, *Heterochrony: The Evolution of Ontogeny* (New York: Plenum, 1991). © 1991, Plenum Publishing Corp.

p. 118: redrawn by permission of the Trustees of the Western Australian Museum from paintings by Martin Thompson in Barry Hutchins and Martin Thompson, *The Marine and Estuarine Fishes of South-western Australia* (Perth: Western Australian Museum, 1995).

p. 122: redrawn from K. J. McNamara, "Sexual Dimorphism: The Role of Heterochrony," in K. J. McNamara, ed., *Evolutionary Change and Heterochrony* (Chichester and New York: Wiley, 1995).

p. 149: redrawn by permission from Jonathan Kingdon, *East African Mammals*, vol. IIIC, *Bovids* (Chicago: University of Chicago Press, 1982).

p. 174: redrawn from B. Swedmark, "The Interstitial Fauna of Marine Sand," *Biological Reviews* 39 (1964): 1–42. © 1964; reprinted with permission of Cambridge University Press.

p. 177: reproduced, with permission of the artist, from D. Korn, "Impact of Environmental Perturbations on Heterochronic Development in Palaeozoic Ammonoids," in K. J. McNamara, ed., *Evolutionary Change and Heterochrony* (Chichester and New York: Wiley, 1995).

p. 188: redrawn by permission of Blackwell Science Pty. Ltd. from Andrew Cockburn, *An Introduction to Evolutionary Biology* (Oxford: Blackwell Scientific Publications, 1992). © 1992, Blackwell Science Pty. Ltd.

p. 201: redrawn by permission from *Nature* 375 (1995): 678–81, P. Sordino et al., "*Hox* Gene Expression in Teleost Fins and the Origin of Vertebrate Digits," © 1995, Macmillan Magazines Ltd.

p. 215: based on paintings by John Gurche in John Reader, *The Rise of Life* (London: Collins, 1986).

p. 216: redrawn by permission of the Paleontological Society from B. J. MacFadden, "Fossil Horses from '*Eohippus*' (*Hyracotherium*) to *Equus*: Scaling, Cope's Law, and the Evolution of Body Size," *Paleobiology* 12 (1986): 355–69.

Index

LIBRARY OF CONGRESS CATALOGING-IN-PUBLICATION DATA

McNamara, Ken.
    Shapes of time : the evolution of growth and development / Kenneth J.
McNamara.
      p. cm.
    Includes bibliographical references and index.
    ISBN 0-8018-5571-3 (alk. paper)
     1. Heterochrony (Biology)   I. Title.
QH395.M36 1997   96-29775
576.8—dc21   CIP